Memoirs of the American Mathematical Society

Number 251

T. A. Chapman

Approximation results in topological manifolds

Published by the

AMERICAN MATHEMATICAL SOCIETY

Providence, Rhode Island, USA

November 1981 · Volume 34 · Number 251 (second of 5 numbers)

MEMOIRS of the American Mathematical Society (ISSN 0065-9266) is published bimonthly (each volume consisting usually of more than one number) by the American Mathematical Society at 201 Charles Street, Providence, Rhode Island 02904. Second Class postage paid at Providence, Rhode Island 02940. Postmaster: Send address changes to Memoirs of the American Mathematical Society, American Mathematical Society, P. O. Box 6248, Providence, RI 02940.

ABSTRACT. A number of results are established concerning the approximation of maps between topological manifolds by homeomorphisms, approximate fibrations, and block bundles. Applications are given to the approximation of approximate fibrations by block bundles, and to the taming of locally homotopically unknotted embeddings.

AMS (MOS) SUBJECT CLASSIFICATIONS (1970). Primary 57A15, 57A30.

KEY WORDS AND PHRASES. Approximate fibrations, block bundles, engulfing, ε-equivalences.

Library of Congress Cataloging in Publication Data

Chapman, T. A. (Thomas A.), 1940-
Approximation results in topological manifolds.

(Memoirs of the American Mathematical Society, ISSN 0065-9266 ; no. 251)
Bibliography: p.
1. Topological manifolds. I. Title. II. Series.
QA3.A57 no. 251 [QA613.2] 510s [514'.3] 81-12791
ISBN 0-8218-2251-9 AACR2

1. INTRODUCTION

In this paper the term *manifold* will be used to mean a finite-dimensional topological m-manifold M^m --possibly with boundary ∂M^m. By a *PL manifold* we will mean a piecewise linear manifold as in [16]. In a very broad sense the purpose of this paper is to investigate questions of the following type: *If* $f: M \to N$ *is a map between manifolds, when can* f *be approximated by "nice" maps?* The "nice" maps referred to include homeomorphisms, approximate fibrations, and block bundles. The key idea common to all of these results is a very carefully controlled engulfing. This is established very early in Section 3 and is then used at crucial points throughout the paper. This entire line of research is a natural outgrowth of the approximation results of [4] and [7]. The statements and proof of the theorems are influenced heavily by the corresponding Q-manifold results of [6].

For the most part all spaces in this paper will be locally compact, separable and metric. A map $f: X \to Y$ (i.e., a continuous function) is *proper* provided that $f^{-1}(C)$ is compact, for all compact $C \subset Y$. If α is an open cover of Y, then a proper map $f: X \to Y$ is said to be an *α-fibration* (or, f has the *α-lifting property*) if for all maps $F: Z \times [0, 1] \to Y$ and $\tilde{F}_0: Z \to X$ for which $f\tilde{F}_0 = F_0$, there is a map $G: Z \times [0, 1] \to X$ such that $G_0 = \tilde{F}_0$ and fG is α-close to F. This latter statement means that given any $(z, t) \in Z \times [0, 1]$ there is a $U \in \alpha$ containing both $fG(z, t)$ and $F(z, t)$. Finally a proper map $p: X \to Y$ is said to be an *approximate fibration* provided that it has the α-lifting property, for all open covers α of Y. We refer the reader to [8], where the notion of an approximate fibration was introduced and some of its basic properties were established.

Supported in part by NSF Grant MCS 76-06929.

Received by the editors May, 1979.

Our first result is concerned with the homotopy detection of those maps which are close to approximate fibrations.

THEOREM 1. *Let* B *be a space which is locally polyhedral, let* α *be an open cover of* B*, and let* $m \geq 5$ *be an integer. There exists an open cover* β *of* B *so that if* M^m *is a manifold* ($\partial M = \emptyset$) *and* $f: M \to B$ *is a* β*-fibration, then* f *is* α*-close to an approximate fibration* $p: M \to B$.

There are two variations of this result which follow from the same proof. The first assumes that $\partial M \neq \emptyset$ and f is already an approximate fibration over an open set in B which contains $f(\partial M)$. The conclusion is that the approximate fibration $p: M \to B$ can be chosen to agree with f on ∂M. The second variation assumes that $f|\partial M: \partial M \to B$ is the projection map of a locally trivial bundle or a block bundle over arbitrarily small simplices. The conclusion is that p can be chosen to agree with f on ∂M.

The proof of Theorem 1, which is given in Section 6, uses a handle lemma from Section 5 that is established by engulfing and torus geometry. The idea of the proof is to construct maps $f_i: M \to B$ which are close to f and which are β_i-fibrations, where $\{\beta_i\}$ is a sequence of open covers of B whose mesh converges to 0. The f_i are additionally chosen so that $\lim f_i = p: M \to B$ is defined and is our desired approximate fibration.

In [15] there is constructed an example of an approximate fibration $p: M^6 \to S^1$ ($\partial M = \emptyset$) whose fiber is not homotopy equivalent to any finite complex. Perhaps the following result will be useful in determining all such fibers (up to homotopy type).

COROLLARY. *A polyhedron* K *is homotopy equivalent to the (homotopy) fiber of some approximate fibration* $p: M^m \to S^1$ ($m \geq 5$ *and* $\partial M = \emptyset$) *if and only if* $K \times S^1$ *is homotopy equivalent to some compact* m*-manifold without boundary.*

The following notation will be used throughout this paper. If α is an open cover of Y, then a homotopy $h_t: X \to Y$ is said to be an α*-homotopy* provided that each set $\{h_t(x) \mid 0 \le t \le 1\}$ lies in some element of α. A proper map $f: X \to Y$ is said to be an *α-equivalence* if there is a proper map $g: Y \to X$ and proper homotopies $\phi_t: gf \simeq id_X$, $\theta_t: fg \simeq id_Y$ such that $f\phi_t: X \to Y$ and $\theta_t: Y \to Y$ are α-homotopies. We write this as

$$\phi_t: gf \overset{f^{-1}(\alpha)}{\simeq} id \qquad \text{and} \qquad \theta_t: fg \overset{\alpha}{\simeq} id,$$

where $f^{-1}(\alpha)$ denotes the open cover of X defined by $f^{-1}(\alpha) = \{f^{-1}(U) \mid U \in \alpha\}$. Finally if Y has a specified metric and $\varepsilon > 0$ is given, then we will also use ε to denote the open cover of Y by balls of diameter $< \varepsilon$. This convention means that we have also defined the notions of ε-homotopy and ε-homotopy equivalence.

It might be of some interest to point out another application of Theorem 1. In [7] the following α-Approximation Theorem was established: *For every manifold N^m, $m \ge 5$, and open cover α of N, there exists an open cover β of N such that if M^m is a manifold and $f: M \to N$ is a β-equivalence which is a homeomorphism from ∂M to ∂N, then f is α-homotopic* rel ∂M *to a homeomorphism.* Here is another proof of this result based on Theorem 1 above and Edwards' Approximation Theorem [9]. For simplicity assume that $\partial M = \emptyset = \partial N$. First by use of Theorem 1 we may assume that f is an approximate fibration . Next it easily follows that f is CE, i.e., the inverse image of each point has property UV^∞ in M. Finally we use Edwards' Approximation Theorem to approximate f by a homeomorphism. One advantage of this proof over the proof given in [7] is that surgery theory is completely avoided, i.e., we do not need the fact that any homotopy $I^i \times T^{m-i}$ is a real $I^i \times T^{m-i}$ rel ∂.

Let F^m be a compact manifold for which the Whitehead group, $Wh(F)$, vanishes. We use $\mathcal{S}(F)$ to denote the set of equivalence classes of the form $[f]$, where $f: M^m \to F$ is a homotopy equivalence of a compact manifold to F which is a homeomorphism from ∂M to ∂F. Another such map, $f': M' \to F$, is defined to be *equivalent* to f provided that there exists a homeomorphism $h: M \to M'$ for which $hf' \simeq f$. Note that via use of the s-cobordism theorem,

$\mathcal{S}(F)$ is just the set of homotopy-topological structures on F rel ∂ [17, p.265]. If T^n is the n-torus and $e: T^n \to T^n$ is any standard finite cover, then there is a transfer map $e^{\#}: \mathcal{S}(T^n \times F) \to \mathcal{S}(T^n \times F)$ defined by $e^{\#}([f]) = [\tilde{f}]$, where $\tilde{f}$ comes from the pullback diagram

$$\begin{array}{ccc} \tilde{M} & \xrightarrow{\tilde{f}} & T^n \times F \\ \downarrow & & \downarrow e \times id \\ M & \xrightarrow{f} & T^n \times F \end{array} .$$

We use $\mathcal{S}_o(T^n \times F)$ to denote those elements of $\mathcal{S}(T^n \times F)$ that are invariant under any of these transfer maps.

Our metric on euclidean n-space R^n is the one derived from the norm $\|x\| = (|x_1|^2 + \cdots + |x_n|^2)^{1/2}$. In the following statement p will denote projection to R^n.

THEOREM 2. *Let* $n \geq 0$ *be an integer. For every* $\varepsilon > 0$ *there exists a* $\delta > 0$ *so that if* $f: M \to R^n \times F$ *is a* $p^{-1}(\delta)$*-equivalence for which* $f|\partial M: \partial M \to R^n \times \partial F$ *is a homeomorphism, where* M^{m+n} *is a manifold,* F^m *is as above and* $m + n \geq 5$, *then there is an element* $\sigma(f)$ *of* $\mathcal{S}_o(T_n \times F)$ *which vanishes if and only if* f *is* $p^{-1}(\varepsilon)$*-homotopic to a homeomorphism.*

The proof of Theorem 2, which is given in Section 8, is purely geometric in nature. It uses only engulfing and torus geometry. The utility of this result is that many times a geometric problem is encountered which gives rise to a $p^{-1}(\delta)$-equivalence $f: M \to R^n \times F$ as above; thus an obstruction is encountered in $\mathcal{S}_o(T^n \times F)$. In certain cases surgery theory tells us that $\mathcal{S}_o(T^n \times F)$ vanishes. For example it follows from [17, p.285] that if F^m is a $K(\pi, 1)$ with π poly Z and $m + n \geq 5$, then $\mathcal{S}(T^n \times F) = 0$. So in particular $S_o(T^n \times F) = 0$. The following result deals with simply connected fibers.[†]

THEOREM 3. *If* F^m *is a compact simply connected manifold and* $m + n \geq 5$, *then* $S_o(T^n \times F)$ *is in* 1-1 *correspondence with the set of homotopy classes*

$$[F_o, \partial F_o; G/TOP, *],$$

where F_o *is the manifold obtained by deleting an interior point of* F.

Recall that G/TOP is the homotopy fiber of the natural map of classifying spaces, $BTOP \to BG$. It is an H-space and the functor $X \mapsto [X; G/TOP]$ is a contravariant functor to abelian groups. The proof of Theorem 3, which is given in Section 9, uses the surgery exact sequence [17, p.269].

Combining Theorem 3 with the remarks made above about fibers F which are $K(\pi, 1)$'s we have the following corollary. It is used in Appendix 2 to show that locally homotopically unknotted embeddings are locally flat.

COROLLARY. *If* F *is a sphere, then* $S_o(T^n \times F) = 0$.

Here is a generalization of the result of [4] on approximating maps into bundles by homeomorphisms.

THEOREM 4. *Let* B^n *be a manifold, let* α *be an open cover of* B, *and let* $m + n \geq 5$ *be an integer. There exists an open cover* β *of* B *so that if* M^{m+n} *is a manifold,* $p\colon E^{m+n} \to B$ *is a fiber bundle with fiber a compact manifold* F, *and* $f\colon M \to E$ *is a* $p^{-1}(\beta)$*-equivalence which is a homeomorphism from* ∂M *to* ∂E, *then* f *is* $p^{-1}(\alpha)$*-homotopic* (rel ∂M) *to a homeomorphism provided that* $S_o(T^i \times I^j \times F) = 0$, *for* $i + j = n$.

The idea of the proof (see Section 10) is to locally work through a handle decomposition of B. Starting with the handles of index 0 we use Theorem 2

[†]The author is indebted to L. C. Siebenmann for providing him with this computation.

as a handle lemma to deform f to a map which is a homeomorphism over these handles. Then we inductively continue through the handles of positive index.

From the remarks following Theorem 2 we conclude that Theorem 4 is true if F is a $K(\pi, 1)$ with π poly Z. We also point out that Theorem 4 is true if we merely assume that $\mathcal{S}_o(T^i \times I^j \times F) = 0$ for j ranging through the indices of a given handle decomposition of B. For example if B is a sphere, it has a handle decomposition with handles of only index 0 and $\dim B$, so we only need $\mathcal{S}_o(T^i \times I^j \times F) = 0$ for $j = 0$ and $j = \dim B$. This is somewhat different from the procedure of [7] or [20], in which the degree of the approximation depended on the size of the handles in the target.

Finally we mention that the techniques of this paper have important applications to the homotopy detection of local flatness and to the homotopy detection of those maps which are close to block bundle maps. More specifically in Appendix 2 we sketch a proof that a locally homotopically unknotted manifold pair is locally flat. This is essentially what appears in [5], but now we capture some extra dimensions that eluded us at the time. Results of this type are not new any longer and we refer the reader to Appendix 2 for a fuller account of the literature on the subject. In Appendix 3 we show how the approximation techniques of this paper can be used to prove that approximate fibrations can be approximated by block bundles provided that a π_1-condition on the fiber is satisfied. Again this is not a new result--it having been done recently by Quinn [18].

The key idea in this paper is an engulfing trick, which might best be described as a "shuffle." To acquaint the reader immediately with this we sketch a quick proof of the α-Approximation Theorem of [7]. This is done in Appendix 1 and we recommend that it be read first before anything else is attempted.

2. PRELIMINARIES

In this section we will introduce some additional notation and remind the reader of some well-known results from the literature which will be used throughout this paper.

Recall from Section 1 that R^n denotes euclidean n-space, i.e., $R^n = R \times \cdots \times R$ (n times). In R^n we consider n-cells of the form $B_r^n = [-r, r]^n$. We use S^1 to denote the set of complex numbers of absolute value 1, and the n-torus is $T^n = S^1 \times S^1 \times \cdots \times S^1$. We always let $I = [0, 1]$, and the standard n-cell is $I^n = [0, 1] \times \cdots \times [0, 1]$.

Expanding on the definition of an α-equivalence which was given in Section 1 let $f: X \to Y$ be proper, let $C \subset Y$ be closed, and let α be an open cover of Y. Then f is said to be an α-*equivalence over* C if there is a proper map $g: C \to X$ and proper homotopies $\phi_t: gf|f^{-1}(C) \overset{f^{-1}(\alpha)}{\simeq} \text{id}$, $\theta_t: fg \overset{\alpha}{\simeq} \text{id}$. This means that $\phi_t: f^{-1}(C) \to X$ and $\theta_t: C \to Y$ are proper homotopies such that $\phi_o = \text{id}$, ϕ_1 is the inclusion $f^{-1}(C) \hookrightarrow X$, ϕ_t is an $f^{-1}(\alpha)$-homotopy, $\theta_o = \text{id}$, θ_1 is the inclusion $C \hookrightarrow Y$, and θ_t is an α-homotopy. Note that if $C \subset G \subset Y$, where C is closed and G is open, then for all sufficiently fine open covers α of Y, any α-equivalence $f: X \to Y$ restricts to an $(\alpha \cap G)$-equivalence over C, $f|f^{-1}(G): f^{-1}(G) \to G$. (Here $\alpha \cap G = \{U \cap G | U \in \alpha\}$.

We write "f = g over C" to mean that $f^{-1}(C) = g^{-1}(C)$ and $f = g$ on $f^{-1}(C)$. In general we say that f has property P over C whenever $f|f^{-1}(C)$ has property P. In some instances when the meaning is clear we will simply write f when the restriction $f|$ is intended. Finally, a homotopy $f_t: X \to Y$ is said to be an α-*homotopy over* $C \subset Y$ if $f_t|f_o^{-1}(C): f_o^{-1}(C) \to Y$ is an α-homotopy.

The following result tells us how to detect α-equivalences locally. It is established in [6].

PROPOSITION 2.1. *Let* B *be a space and let* γ *be an open cover of* B. *For every open cover* α *of* B *there exists an open cover* β *of* B *so that if* $f: X \to Y$ *and* $p: Y \to B$ *are proper maps such that* X, Y *are* ANRs *and* f *is a* $p^{-1}(\beta)$*-equivalence over the closure of each element of* γ*, then* f *is a* $p^{-1}(\alpha)$*-equivalence.*

REMARKS. 1. There is a similarly-worded local version of this result whose conclusion states that f is a $p^{-1}(\alpha)$-equivalence over some closed set $C \subset Y$.
2. The above result is also true if we merely assume that X, Y are separable metric ANRs and f, p are no longer necessarily proper. For this to make sense we also have to drop the requirement that the map g and homotopies ϕ_t, θ_t of the definition are proper.

We now expand on the definition of an α-fibration which was given in Section 1. Let $f: X \to Y$ be proper, let $C \subset Y$ be closed, and let α be an open cover of Y. Then f is said to be an α*-fibration over* C if for all maps $F: Z \times I \to C$ and $\tilde{F}_0: Z \to X$ for which $f\tilde{F}_0 = F_0$, there is a map $G: Z \times I \to X$ such that $G_0 = \tilde{F}_0$ and fG is α-close to F. Note that if $f: X \to Y$ is an α-fibration and $C \subset G \subset Y$, where C is closed and G is open, then the restriction $f|f^{-1}(G): f^{-1}(G) \to G$ is an $(\alpha \cap G)$-fibration over C provided that α is sufficiently fine.

Here is an analogue of Proposition 2.1 for α-fibrations. Again see [6] for a proof.

PROPOSITION 2.2. *Let* B *be an* ANR *and let* γ *be an open cover of* B. *For every open cover* α *of* B *there exists an open cover* β *of* B *so that if* X *is an* ANR *and* $f: X \to B$ *is a* β*-fibration over the closure of each element of* γ*, then* f *is an* α*-fibration.*

REMARK. As in Proposition 2.1 there is a similarly-worded local version of this result whose conclusion states that f is an α-fibration over some closed set $C \subset B$.

The following result gives us another point of view from which we can recognize $p^{-1}(\alpha)$-equivalences. See [6] for a proof.

PROPOSITION 2.3. *Let* B *be an* ANR *and let* α *be an open cover of* B. *There exists an open cover* β *of* B *so that if* $p\colon E \to B$ *is a Hurewicz fibration,* X *is arbitrary, and* $f\colon X \to E$ *is a homotopy equivalence for which* $pf\colon X \to B$ *is a* β*-fibration, then* f *is a* $p^{-1}(\alpha)$*-equivalence.*

3. ENGULFING

In this section we will establish Theorem 3.6, which is the main engulfing result which will be needed in the sequel. It is deduced rather formally from an engulfing result of [22] which is stated below in Lemma 3.1. We will first need a definition. A polyhedron P in a manifold M is said to be *locally polyhedral in* M if for each $x \in P$ there is an open set $U \subset M$ of x and a triangulation of U as a PL manifold so that $U \cap P$ is a subpolyhedron of this triangulation. Here is Proposition 2.3 of [22].

LEMMA 3.1. *Let* M^m *be a manifold,* $\partial M = \emptyset$, *let* $p \geq 0$ *be an integer, let* $P \subset M$ *be a polyhedron which is closed and locally polyhedral in* M, $\dim P \leq m - 3$, *and let* $P_o \subset P$ *be a subpolyhedron so that* $Q = \overline{P - P_o}$ *is compact and* $\dim Q \leq p$. *Also let* $U_o \subset U_1 \subset \cdots \subset U_p$ *and* $M_o \subset M_1 \subset \cdots \subset M_p = M$ *be non-null open subsets of* M *such that* $P \subset M_o$, $P_o \subset U_o$, $U_i \subset M_i$, *and such that for* $0 \leq i \leq p - 1$, *all maps* $(K, L) \to (M_i, U_i)$ *of a finite simplicial pair of dimension* $\leq p - i$ *are homotopic in* (M_{i+1}, U_i) *to a map* $(K, L) \to (U_{i+1}, U_i)$. *Then there exists an isotopy, with compact support, of* id_M *to a homeomorphism* $h\colon M \to M$ *for which* $P \subset h(U_p)$.

We now derive an easy consequence of Lemma 3.1 which will be more directly useful to us. For notation let $u, v\colon [0, \infty) \to (-2, 2) \subset R$ be maps so that we have $v(s) < u(s)$, for all $s \geq 2$.

We also use $\Gamma(u)$ and $\Gamma(v)$ to denote the graphs under u and v. That is, $\Gamma(u)$ and $\Gamma(v)$ are the subsets of $[0, \infty) \times R$ defined by

$$\Gamma(u) = \{(s, t) \mid -\infty < t \leq u(s)\},$$
$$\Gamma(v) = \{(s, t) \mid -\infty < t \leq v(s)\}.$$

LEMMA 3.2. *For each manifold* M^m, $\partial M = \emptyset$, *there exists an* $\varepsilon = \varepsilon(m) > 0$ *so that if* $f: M \to [0, \infty) \times R$ *is an* ε-*fibration over* $[0, 4] \times [-4, 4]$, *and* $P \subset M$ *is a polyhedron which is closed and locally polyhedral in* M, $\dim P \leq m - 3$, *then there exists an isotopy, which is supported on* $f^{-1}([0,3] \times [-3,3])$, *of* id_M *to a homeomorphism* $h: M \to M$ *which satisfies* $P \cap f^{-1}(\Gamma(v)) \subset hf^{-1}(\Gamma(u))$.

Proof. Without loss of generality assume that $P \subset f^{-1}(\Gamma(v))$. Choose a closed subpolyhedron P_o of P in $f^{-1}([2, \infty) \times R)$ so that $P \cap f^{-1}([2.5, \infty) \times R)$ lies in P_o. Choose continuous functions $u_i: [0, \infty) \to (-2, 2)$, $v_i: [0, \infty) \to (-2, 2)$, $0 \leq i \leq m$, so that

(1) $v(s) < u_o(s) < v_o(s) < \cdots < u_m(s) < v_m(s)$, for all $s \geq 2$,

(2) $u_o < u_1 < \cdots < u_m = u$,

(3) $v < v_o < \cdots < v_m$,

(4) $u_i < v_i$, for all i.

It should be pointed out that the choice of the u_i and v_i depend only on u and v, and the ε must be calculated in terms of these choices. We are assuming that f is an ε-fibration over $[0, 4] \times [-4, 4]$, thus for each i there is a homotopy of the identity on $(f^{-1}(\Gamma(v_i)), f^{-1}(\Gamma(u_i)))$ to a mapping into $(f^{-1}(\Gamma(u_{i+1})), f^{-1}(\Gamma(u_i)))$, with the homotopy taking place in $(f^{-1}(\Gamma(v_{i+1})), f^{-1}(\Gamma(u_i)))$. Also we have $P_o \subset f^{-1}(\Gamma(u_o))$. Now a quick application of Lemma 3.1 gives our desired homeomorphism $h: M \to M$. □

The following result is a generalization of Lemma 3.2. Its proof is based on the proof of Theorem 2.1 of [22]. For notation we still have the maps $u, v: [0, \infty) \to (-2, 2)$ of Lemma 3.2.

LEMMA 3.3. *For each manifold* M^m, $\partial M = \emptyset$ *and* $m \geq 5$, *there exists an* $\varepsilon = \varepsilon(m) > 0$ *so that if* $f: M \to [0, \infty) \times R$ *is an* ε-*fibration over* $[0,4] \times [-4,4]$, *then there is an isotopy, which is supported on* $f^{-1}([0, 3] \times [-3, 3])$, *of* id_M *to a homeomorphism* $h: M \to M$ *which satisfies* $f^{-1}(\Gamma(v)) \subset h(f^{-1}(\Gamma(u))$.

Proof. Choose maps $v_1, v_2: [0, \infty) \to (-2, 2)$ so that $v < v_1 < v_2$ and $v_2(s) < u(s)$ for all $s \geq 2$. If $P^{m-3} \subset M$ is a polyhedron which is closed and locally polyhedral, then by Lemma 3.2 there is a homeomorphism $h_1: M \to M$ which is supported on $f^{-1}([0, 2] \times [-2, 2])$ and for which $P \cap f^{-1}(\Gamma(v_2))$ lies in $h_1(f^{-1}(\Gamma(u))$. If $Q^2 \subset M - P$ is a polyhedron which is closed and locally polyhedral in M, then Lemma 3.2 again provides us with a homeomorphism $h_2: M \to M$ which is supported on $f^{-1}(\Gamma(v_1) \cap ([0, 3] \times [-3, 3]))$ and for which $h_2(Q \cap f^{-1}([0, 2.1] \times [-2.1, \infty))$ lies in $f^{-1}(\Gamma(v))$.

According to Proposition 2.7 of [22] we can choose P and Q to be "topological dual skeleta" in M. This means that for every $\delta > 0$ we can choose P and Q so that if $C \subset M - P$ is a compactum, then there are homeomorphisms of M, with compact support and which are δ-close to id_M, which take C as close to Q as we want. The compactum that we have in mind is the closure of

$$h_1(M - f^{-1}(\Gamma(u))) \cap f^{-1}(\Gamma(v_2)),$$

which we call C. Let $h_3: M \to M$ be a homeomorphism which is supported on $f^{-1}([0, 3] \times [-3, 3])$, which is δ-close to id_M, and which takes C close to Q. It is easy to see that with an appropriate choice of δ, $h = h_2 h_3 h_1$ fulfills our requirements. □

We now deduce the following "radial" engulfing result from Lemma 3.3. For notation B will be a finite-dimensional polyhedron which will act as a parameter space. Also B will have the metric topology determined by a fixed triangulation, and $B \times R$ will have the metric which is the product of the metric on B with the standard metric on R.

LEMMA 3.4. *For every integer* $m \geq 5$ *and* $\varepsilon > 0$ *there exists a* $\delta > 0$ *so that if* M^m *is a manifold,* $\partial M = \emptyset$, *and* $f: M \to B \times R$ *is a* δ-*fibration over* $B \times [-4, 4]$, *then there exists a homeomorphism* $h: M \to M$ *which is supported on* $f^{-1}(B \times [-3, 3])$ *and which satisfies* $f^{-1}(B \times (-\infty, 1]) \subset hf^{-1}(B \times (-\infty, 0))$.

Moreover if $p_B = \text{proj}: B \times R \to B$, *then there is an isotopy* $h_t: id_M \simeq h$ *which is supported on* $f^{-1}(B \times [-3, 3])$ *and which is a* $(p_B f)^{-1}(\varepsilon)$*-homotopy.*

Proof. Choose a fine triangulation of B whose mesh depends on ε. Once this is done the δ is calculated so that the following constructions make sense. For each vertex v of B let C_v be a small closed neighborhood of v and let $\tilde{C}_v$ be an open set containing C_v so that the $\tilde{C}_v$'s are pairwise-disjoint. We are going to apply Lemma 3.3 to each composition

$$(q \times id)f: f^{-1}(\tilde{C}_v \times R) \xrightarrow{f} \tilde{C}_v \times R \xrightarrow{q\times id} [0, \infty) \times R,$$

where $q: \tilde{C}_v \to [0, \infty)$ is a proper map. To see how Lemma 3.3 applies first note that we may take $q: \tilde{C}_v \times [0, \infty)$ to be a proper retraction. We can choose δ small enough so that $f: f^{-1}(\tilde{C}_v \times R) \to \tilde{C}_v \times R$ is a δ-fibration over $q^{-1}([0, 4] \times [-4, 4])$, yet $(q \times id)f: f^{-1}(\tilde{C}_v \times R) \to [0, \infty) \times R$ might not be a δ-fibration over $[0, 4] \times [-4, 4]$. However in the proof of Lemma 3.3 we did not need the full strength of the ε-lifting property over $[0, 4] \times [-4, 4]$. We only needed the ε-lifting property for homotopies which move only in the R-direction. It is easy to see that this is true for the map $(q \times id)f$: $f^{-1}(\tilde{C}_v \times R) \to \tilde{C}_v \times R$. Now applying Lemma 3.3 to $(q \times id)f: f^{-1}(\tilde{C}_v \times R) \to \tilde{C}_v \times R$ we obtain a homeomorphism $h(v): M \to M$ which is supported on $f^{-1}(\tilde{C}_v \times [-2, 2])$ and for which $f^{-1}(C_v \times (-\infty, 1.5]) \subset h(v)f^{-1}(\tilde{C}_v \times (-\infty, 0))$. The $h(v)$'s then compose to yield a homeomorphism $h^o: M \to M$ such that

$$f^{-1}((\cup C_v) \times (-\infty, 1.5]) \subset h^o f^{-1}(B \times (-\infty, 0)).$$

This completes the first step of the construction.

The effect of h^o was to deal with the 0-skeleton. The next step is to show how to deal with the 1-skeleton. Let σ be a 1-simplex in B with vertices v_1, v_2, and let C_σ be a closed set containing the closure of $\sigma - (C_{v_1} \cup C_{v_2})$ in its interior. Let $\tilde{C}_\sigma$ be a slightly larger open set containing C_σ. This can be done so that $\tilde{C}_\sigma$'s are pairwise disjoint and so that $\tilde{C}_\sigma$ lies in a small neighborhood of σ. Again using Lemma 3.3 there is a homeomorphism $h(\sigma): M \to M$ which is supported on $f^{-1}(\tilde{C}_\sigma \times [-2.5, 2.5])$ and for which $f^{-1}((C_\sigma \cup [\tilde{C}_\sigma \cap (C'_{v_1} \cup C'_{v_2})]) \times (-\infty, 1.4])$ lies in

$$h(\sigma)f^{-1}[(\tilde{C}_\sigma \times (-\infty, -2)) \cup ((\tilde{C}_\sigma \cap (C_{v_1} \cup C_{v_2})) \times (-\infty, 1.5))],$$

where C'_{v_1} and C'_{v_2} are slightly smaller neighborhoods of v_1 and v_2, respectively. Do this for each σ and then compose the $h(\sigma)$'s to obtain a homeomorphism $h^1: M \to M$. We note that $h^1h^of^{-1}(B \times (-\infty, 0))$ contains $f^{-1}(B^1 \times (-\infty, 1.4])$, where B^1 is the 1-skeleton of B. If we continue to work through the skeleta of B in this manner we eventually obtain our desired homeomorphism $h: M \to M$. The $(p_Bf)^{-1}(\varepsilon)$-homotopy $h_t: \mathrm{id} \simeq h$ is clear from the construction because $\mathrm{diam}(\Delta) << \varepsilon$, for all simplices Δ in the triangulation of B. □

We now generalize Lemma 3.4. For notation B will still be a finite dimensional polyhedron which acts as a parameter space and $\theta: R \to R$ will be a homeomorphism which is supported on $[-1, 1]$.

LEMMA 3.5. *For every integer* $m \geq 5$ *and* $\varepsilon > 0$ *there exists a* $\delta > 0$ *so that if* M^m *is a manifold,* $\partial M = \emptyset$, *and* $f: M \to B \times R$ *is a* δ*-fibration over* $B \times [-4, 4]$, *then there exists a homeomorphism* $\tilde{\theta}: M \to M$ *which is supported on* $f^{-1}(B \times [-2, 2])$ *and for which* $d(f\tilde{\theta}, (\mathrm{id}_B \times \theta)f) < \varepsilon$. *Moreover we can construct* $\tilde{\theta}$ *so that there is a* $(p_Bf)^{-1}(\varepsilon)$*-isotopy of* $\tilde{\theta}$ *to* id *which is supported on* $f^{-1}(B \times [-2, 2])$, *where* $p_B = \mathrm{proj}: B \times R \to B$.

Proof. Choose a fine partition of $[-1, 1]$, $-1 = x_o < x_1 < \cdots < x_{n-1} < x_n = 1$ By stacking together the homeomorphisms of Lemma 3.4 we will construct a homeomorphism $\tilde{\theta}: M \to M$ which is supported on $f^{-1}(B \times [-2, 2])$, which satisfies $d(p_Bf\tilde{\theta}, p_Bf) < \varepsilon/2$, and which also satisfies

$$f^{-1}(B \times (-\infty, \theta(x_{i-1})]) \subset \tilde{\theta}f^{-1}(B \times (-\infty, x_i]) \subset f^{-1}(B \times (-\infty, \theta(x_i)]),$$

for $1 \leq i \leq n - 1$. If the support of $\tilde{\theta}$ is sufficiently close to

$f^{-1}(B \times [-1, 1])$, and if the partition $\{x_i\}$ is sufficiently fine, then we will clearly have $d(f\tilde{\theta}, (\mathrm{id} \times \theta)f) < \varepsilon$ as desired. The isotopy of $\tilde{\theta}$ to id will be clear from the construction. We will actually construct $\tilde{\theta}$ as a composition, $\tilde{\theta} = \tilde{\theta}_{n-1} \circ \cdots \circ \tilde{\theta}_1$. The working procedure is to assume that δ is small enough to perform all of the ensuing constructions. Just to make sure that there is no confusion, here is the order of choices: First the partition $\{x_i\}$ is chosen, then δ is calculated in terms of this choice.

We start by constructing $\tilde{\theta}_1$. By Lemma 3.4 there is a homeomorphism $\tilde{\theta}_1$: $M \to M$ which is supported on $f^{-1}(B \times [-1, 1])$, for which $d(p_B f\tilde{\theta}_1, p_B f)$ is small, and for which

$$(*) \quad f^{-1}(B \times (-\infty, 1]) \subset \tilde{\theta}_1 f^{-1}(B \times (-\infty, x_1]) \subset f^{-1}(B \times (-\infty, \theta(x_1)]).$$

Note that the first inclusion will be automatically satisfied provided that $\tilde{\theta}_1$ is supported on $f^{-1}(B \times [-1, 1])$.

We will construct $\tilde{\theta}_2$ as a composition, $\tilde{\theta}_2 = \theta_2'' \circ \theta_2'$. It will be supported on $\tilde{\theta}_1 f^{-1}(B \times [x_1, 1])$, so the inclusion (*) will be automatically preserved. The homeomorphism θ_2' comes from Lemma 3.4. It is supported on $\tilde{\theta}_1 f^{-1}(B \times [x_1, 1])$ so that $d(p_B f\theta_2', p_B f)$ is small and so that

$$f^{-1}(B \times (-\infty, \theta(x_1)]) \subset \theta_2' \tilde{\theta}_1 f^{-1}(B \times (-\infty, x_2]).$$

The homeomorphism θ_2'' also comes from Lemma 3.4. It is supported on $f^{-1}(B \times [\theta(x_1), 1])$ (thus preserving the above inclusion) so that $d(p_B f\theta_2'', p_B f)$ is small and so that

$$\theta_2'' \theta_2' \tilde{\theta}_1 f^{-1}(B \times (-\infty, x_2]) \subset f^{-1}(B \times (-\infty, \theta(x_2)]).$$

Putting together these two inclusions we have

$$f^{-1}(B \times (-\infty, \theta(x_1)]) \subset \theta_2'' \theta_2' \tilde{\theta}_1 f^{-1}(B \times (-\infty, x_2]) \subset f^{-1}(B \times (-\infty, \theta(x_2)]).$$

Therefore $\tilde{\theta}_2 = \theta_2'' \circ \theta_2'$ fulfulls our requirements.

It is now clear that we can inductively continue this process to construct our desired $\tilde{\theta}_i$, $1 \leq i \leq n - 1$. □

We are now ready for our main result. For notation let B be a polyhedron, let $\phi: B \to [0, \infty)$ be proper, and for each $t \in [0, \infty)$ let $B_t = \phi^{-1}([0, t])$. Let $\theta_t: R \to R$ be an isotopy, $t \in [0, \infty)$, which is supported on $[-1, 1]$ and for which $\theta_t = \text{id}$ for $t \geq 1$. This induces a homeomorphism $\theta: B \times R \to B \times R$ defined by $\theta(x, r) = (x, \theta_{\phi(x)}(r))$. It is supported on $B_1 \times [-1, 1]$.

THEOREM 3.6. *For every integer* $m \geq 5$ *and* $\varepsilon > 0$ *there exists a* $\delta > 0$ *such that if* M^m *is a manifold,* $\partial M = \emptyset$, *and* $f: M \to B \times R$ *is a* δ*-fibration over* $B_4 \times [-4, 4]$, *then there exists a homeomorphism* $\tilde{\theta}: M \to M$ *which is supported on* $f^{-1}(B_2 \times [-2, 2])$ *and which satisfies* $d(f\tilde{\theta}, \theta f) < \varepsilon$. *Moreover for every* $\mu > 0$ *there exists a* $\nu > 0$ *such that if* f *is additionally given to be a* ν*-fibration over* $(B_4 - \mathring{B}_{1/3}) \times [-4, 4]$, *then the homeomorphism* $\tilde{\theta}: M \to M$ *additionally satisfies* $d(f\tilde{\theta}, \theta f) < \mu$ *over* $(B - \mathring{B}_{2/3}) \times R$. *Also there is an isotopy of* $\tilde{\theta}$ *to* id *which is supported on* $f^{-1}(B_2 \times [-2, 2])$, *which is a* $(p_B f)^{-1}(\varepsilon)$*-homotopy (globally) and which is a* $(p_B f)^{-1}(\mu)$*-homotopy over* $(B - \mathring{B}_{2/3}) \times R$.

REMARKS ON PROOF. This is similar to the proof of Lemma 3.5. First one establishes a basic engulfing result which is similar to Lemma 3.4. Then imitating the proof of Lemma 3.5 these basic engulfing moves are used to produce our desired homeomorphism $\tilde{\theta}$ satisfying $d(f\tilde{\theta}, \theta f) < \varepsilon$. To achieve the second half of Theorem 3.6 all one has to do is use the ν-control on the homotopies over $(B_4 - \mathring{B}_{1/3}) \times [-4, 4]$ to further refine this process so that we have $d(f\tilde{\theta}, \theta f) < \mu$ over $(B - \mathring{B}_{2/3}) \times R$. □

CONCLUDING REMARKS. Note that in our main results, Lemma 3.5 and Theorem 3.6, we have insisted that $\partial M = \emptyset$. There are also relative versions of these results for M with boundary provided that the hypotheses are strengthened to include a statement that the required engulfing homeomorphisms exist on ∂M. For example in Lemma 3.5 we could additionally assume that $f|\partial M: \partial M \to B \times R$ is the projection map of a locally trivial bundle. Clearly there exists a homeomorphism $\tilde{\theta}_\partial: \partial M \to \partial M$ for which $f\tilde{\theta}_\partial = (\mathrm{id} \times \theta)f|\partial M$, and the homeomorphism $\tilde{\theta}: M \to M$ of Lemma 3.5 can be constructed so that it agrees with $\tilde{\theta}_\partial$ on ∂M. An alternative would be to merely assume that $f|\partial M: \partial M \to B \times R$ is the projection map of a block bundle over simplices of small diameter. This block bundle structure enables one to construct homeomorphisms $\tilde{\theta}_\partial: \partial M \to \partial M$ for which $f\tilde{\theta}_\partial$ is close to $(\mathrm{id} \times \theta)f|\partial M$, and then the procedure is the same as in the bundle case. This is what is done in the proof of the Handle Lemma in Appendix 3.

4. WRAPPING UP

In this section we will establish some results on wrapping up maps around S^1. Our main result is Theorem 4.3, with a simpler (global) version first being established in Lemma 4.1. It is concerned with the problem of wrapping up δ-fibrations around S^1. Theorem 4.2 is the analogue of Lemma 4.1 for wrapping up $p^{-1}(\delta)$-equivalences around S^1. For notation let $e: R \to S^1$ be the covering projection defined by $e(x) = \exp(\pi i\, x/4)$ and let B be a finite dimensional polyhedron.

LEMMA 4.1. *For every integer* $m \geq 5$ *and* $\varepsilon > 0$ *there exists a* $\delta > 0$ *such that if* M^m *is a manifold,* $\partial M = \emptyset$, *and* $f: M \to B \times R$ *is a* δ*-fibration over* $B \times [-4, 4]$, *then there is a manifold* $\tilde{M}$, $\partial\tilde{M} = \emptyset$, *a map* $\tilde{f}: \tilde{M} \to B \times S^1$ *which is an* ε*-fibration, and an open embedding* $\phi: f^{-1}(B \times (-1, 1)) \to \tilde{M}$ *for which the following diagram commutes:*

$$\begin{array}{ccc} \tilde{M} & \xrightarrow{\tilde{f}} & B \times S^1 \\ \uparrow{\scriptstyle \phi} & & \uparrow{\scriptstyle \mathrm{id} \times e} \\ f^{-1}(B \times (-1, 1)) & \xrightarrow{f} & B \times (-1, 1) \end{array}$$

Proof. The proof is exactly like the proof of Lemma 4.1 of [6] which treats the Q-manifold case. The construction of $\tilde{M}$ and $\tilde{f}$ will be important to us in the sequel, so we will recall some of the details of the proof from [6]. We begin by constructing $\tilde{M}$. By Lemma 3.5 we can find an isotopy $h_t: \mathrm{id}_M \simeq h_t$ which is supported on $f^{-1}(B \times [-2.5, 2.5])$, such that $p_B f h_t$ is close to $p_B f$, and such that h_1 is a homeomorphism for which $p_R f h_1 f^{-1}(B \times \{-t\})$ is close to $\{t + 4\}$, for $1.8 \leq t \leq 2.2$ (here p_R denotes projection to R). Let

$$Y = h_1 f^{-1}(B \times (-\infty, -2]) - f^{-1}(B \times (-\infty, -2)),$$

which is a subset of M. Define $\tilde{M} = Y/\sim$, where $\sim$ is the equivalence relation on Y generated by $x \sim h_1(x)$, for all $x \in f^{-1}(B \times \{-2\})$. It is clear that $\tilde{M}$ is a manifold.

We now define the map $\tilde{f}: \tilde{M} \sim B \times S^1$. We can represent S^1 as $[-2, 2]/\sim$, where $\sim$ is generated by $-2 \sim 2$. Thus $f^{-1}(B \times (-1, 1))$ is naturally identified as an open subset of $\tilde{M}$ and $(-1, 1)$ is naturally identified as an open subset of S^1. So we may regard ϕ and $e|(-1, 1)$ as inclusion maps in the commutative rectangle above. Now define a map $g: Y \to B \times [-2, 2]$ as follows. The R-component of g, $p_R g$, sends $f^{-1}(B \times \{-2\})$ to -2, it sends $h_1 f^{-1}(B \times \{-2\})$ to 2, and otherwise it is close to $p_R f|Y$. For the definition of $p_B g: Y \to B$ we will need the inverse isotopy $h_t^{-1}: \mathrm{id} \simeq h_1^{-1}$. Let $\mu: Y \to [0, 1]$ be a Urysohn function which is 0 on $f^{-1}(B \times \{-2\})$ and 1 on $h_1 f^{-1}(B \times \{-2\})$. Define $p_B g(x) = p_B f h^{-1}_{\mu(x)}(x)$, for all $x \in Y$. Note that $g = f$ on $f^{-1}(B \times \{-2\})$ and $p_B g = p_B f h_1^{-1}$ on $h_1 f^{-1}(B \times \{-2\})$. Thus g factors through the appropriate equivalence relations to yield a map $\tilde{f}: \tilde{M} \to B \times S^1$ which makes the above rectangle commute. It is shown in [6] that $\tilde{f}$ is an ε-fibration. □

We are now ready for our analogue of Lemma 4.1 for δ-equivalences. In the following statement B will be a finite dimensional polyhedron, p will denote projection to $B \times R$, and q will denote projection to $B \times S^1$. We will also not require that f and $\tilde{f}$ be proper, but pf and $q\tilde{f}$ will be.

THEOREM 4.2. *For every integer* $m \geq 5$, *and* $\varepsilon > 0$ *there exists a* $\delta > 0$ *such that if* M^m *is a manifold,* $\partial M = \emptyset$, F *is any* ANR, *and* $f: M \to B \times F \times R$ *is a* $p^{-1}(\delta)$*-equivalence over* $B \times F \times [-4, 4]$, *then there is a manifold* $\tilde{M}$, $\partial\tilde{M} = \emptyset$, *a* $q^{-1}(\varepsilon)$*-equivalence* $\tilde{f}: \tilde{M} \to B \times F \times S^1$, *and an open embedding* $\phi: f^{-1}(B \times F \times (-1, 1)) \to \tilde{M}$ *for which the following diagram commutes:*

$$\begin{array}{ccc} \tilde{M} & \xrightarrow{\tilde{f}} & B \times F \times S^1 \\ \uparrow{\scriptstyle \phi} & & \uparrow{\scriptstyle \mathrm{id} \times e} \\ f^{-1}(B \times F \times (-1, 1)) & \xrightarrow{f} & B \times F \times (-1, 1) \end{array}$$

REMARKS ON PROOF. This goes almost word-for-word like the proof of Lemma 4.1. □

The following notation will be needed for Theorem 4.3. Let B be a polyhedron and let $\phi: B \to [0, \infty)$ be a proper map. For each $t \in [0, \infty)$ let $B_t = \phi^{-1}([0, t])$, which is a compactum in B.

THEOREM 4.3. *For every integer* $m \geq 5$ *and* $\varepsilon > 0$ *there exists a* $\delta > 0$ *such that if* M^m *is a manifold,* $\partial M = \emptyset$, *and* $f: M \to B \times R$ *is a* δ*-fibration over* $B_4 \times [-4, 4]$, *then there exists a manifold* $\widetilde{M}$, $\partial\widetilde{M} = \emptyset$, *a map* $\widetilde{f}: \widetilde{M} \to \mathring{B}_3 \times S^1$ *which is an* ε*-fibration over* $B_2 \times S^1$, *and an open embedding* $\phi: f^{-1}(\mathring{B}_1 \times (-1,1)) \to \widetilde{M}$ *so that the following diagram commutes:*

$$\begin{array}{ccc} \widetilde{M} & \xrightarrow{\ \widetilde{f}\ } & \mathring{B}_3 \times S^1 \\ \uparrow{\scriptstyle\phi} & & \uparrow{\scriptstyle \mathrm{id} \times e} \\ f^{-1}(\mathring{B}_1 \times (-1, 1)) & \xrightarrow{\ f\ } & \mathring{B}_1 \times (-1, 1) \end{array}$$

Moreover for every $\mu > 0$ *there exists a* $\nu > 0$ *such that if* f *is additionally given to be a* ν*-fibration over* $(B_4 - \mathring{B}_{1/3}) \times [-4, 4]$, *then the map* $\widetilde{f}: \widetilde{M} \to \mathring{B}_3 \times S^1$ *is additionally a* μ*-fibration over* $(B_2 - \mathring{B}_{2/3}) \times S^1$.

REMARKS ON PROOF. This is similar to the proof of Lemma 4.1. The only significant change that must be made is that Theorem 3.6 is used in lieu of Lemma 3.5. □

CONCLUDING REMARKS. As might be expected from the Concluding Remarks of Section 3, there are also relative versions of these results for manifolds with boundary provided that extra hypotheses are included. Rather than giving specifics here we refer the reader to Lemma 7.4, which is a version of Theorem 4.2 with ∂, and to the proof of the Handle Lemma in Appendix 3, in which a version of Lemma 4.1 with ∂ is used.

5. A HANDLE LEMMA

The purpose of this section is to establish a result which is best described as a handle lemma for Theorem 1. Continuing the strategy of Sections 3 and 4 we first establish the global version (Lemma 5.1) and then quickly obtain the more general version (Theorem 5.2) as a corollary of the proof.

LEMMA 5.1. *If* $\varepsilon > 0$, $m \geq 5$ *and* $n > 0$ *are given, then there exists a* $\delta > 0$ *so that for every manifold* M^m, $\partial M = \emptyset$, $\mu > 0$ *and map* $f: M \to R^n$ *which is a* δ*-fibration over* B_3^n, *there is a map* $\tilde{f}: M \to R^n$ *which is a* μ*-fibration over* B_1^n, *and which is* ε*-homotopic to* f rel $M - f^{-1}(\mathring{B}_3^n)$.

Proof. We begin by introducing some notation. Let $e: R \to S^1$ be the covering projection defined in Section 4 and let $e^n = e \times \cdots \times e: R^n \to T^n$ be the product covering projection. Thus $e^n | B_3^n: B_3^n \to T^n$ is an embedding. We regard $T^{n-1} \times R$ as an open subset of $\mathring{B}_3^n$ so that $e^{n-1} \times \mathrm{id}: B_2^n \to B_2^n$ is the identity (see [10, §8]).

Now consider the restriction

$$f|: f^{-1}(T^{n-1} \times R) \to T^{n-1} \times R.$$

If δ is small enough, then $f|$ is a δ-fibration over $T^{n-1} \times [-3, 3]$. By Lemma 4.1 there is a compact manifold M', a map $f': M' \to T^n$ which is a δ'-fibration, and an open embedding $\phi: f^{-1}(T^{n-1} \times (-2, 2)) \to M'$ so that the following diagram commutes:

$$\begin{array}{ccc} M' & \xrightarrow{\quad f' \quad} & T^n \\ \uparrow & & \uparrow {\scriptstyle \mathrm{id} \times e} \\ f^{-1}(T^{n-1} \times (-2, 2)) & \xrightarrow{\quad f \quad} & T^{n-1} \times (-2, 2) \end{array}$$

Recall that the size of δ' depends on the size of δ. In the remainder of the proof it will be convenient to regard ϕ as an inclusion mapping.

We now unwrap all of this. By passing to covering spaces we obtain a map $f'' \colon M'' \to R^n$ which makes the following pull-back diagram commute:

$$\begin{array}{ccc} M'' & \xrightarrow{f''} & R^n \\ {\scriptstyle p}\downarrow & & \downarrow{\scriptstyle e^n} \\ M' & \xrightarrow{f'} & T^n \end{array}$$

M'' is defined to be the subset of $M' \times R^n$ consisting of all pairs (x, y) for which $f'(x) = e''(y)$. The maps p and f'' are restrictions of projection maps.

ASSERTION. M'' *is a m-manifold and we can choose notation so that* $f'' = f$ *over* $\mathring{B}^n_2$. *Also* f'' *is a* δ''*-fibration (globally), where the size of* δ'' *depends on the size of* δ'.

Proof. Since $e^n \colon R^n \to T^n$ is an immersion (i.e., locally an open embedding) and this is a pullback diagram, it follows that the vertical map $p \colon M'' \to M'$ is also an immersion. Thus M'' is a m-manifold. We chose $e^{n-1} \times \mathrm{id} \colon B^n_2 \to B^n_2$ to be the identity, and from this it easily follows that $p \colon (f'')^{-1}(\mathring{B}^n_2) \to f^{-1}(\mathring{B}^n_2)$ is a homeomorphism. Thus we can choose notation so that $f'' = f$ over $\mathring{B}^n_2$.

To see that f'' is a δ''-fibration let $F \colon Z \times I \to R^n$ and $\tilde{F}_o \colon Z \to M''$ be maps for which $f''\tilde{F}_o = F_o$. Then $e^nF \colon Z \times I \to T^n$ and $p\tilde{F}_o \colon Z \to M'$ are maps for which $f'p\tilde{F}_o = e^nF_o$. Since f' is a δ'-fibration, there is a map $H \colon Z \times I \to M'$ such that $H_o = p\tilde{F}_o$ and $f'H$ is δ'-close to e^nF. Using the fact that $p \colon M'' \to M'$ is a fibration we can lift H to $G \colon Z \times I \to M''$ so that $G_o = \tilde{F}_o$. Thus f'' is a δ''-fibration. Alternately one could also invoke Proposition 2.2 to conclude that f'' is a δ''-fibration. □

We now modify the map f''. Choose $K > 0$ large and let $\gamma \colon R^n \to R^n$ be a radially-defined homeomorphism which takes B^n_K to B^n_3 and keeps B^n_2 pointwise fixed. Now define $\bar{f} = \gamma f'' \colon M'' \to R^n$. Clearly $\bar{f} = f$ over $\mathring{B}^n_2$, $\bar{f}$ is a δ''-fibration, and $\bar{f}$ is a μ-fibration over $R^n - \mathring{B}^n_3$, where the size of μ depends on the size of $1/K$.

The next step is to modify the map $\bar{f}$. Let $\theta\colon R^n \to R^n$ be a homeomorphism which is supported on B^n_9 and which is defined so that

$$\theta^{-1}(x_1, x_2, \ldots, x_n) = (x_1 - 6, x_2, \ldots, x_n)$$

on B^n_2. This can be done so that θ only affects the first coordinate of R^n. Note that $\theta(B^n_2) \subset R^n - \mathring{B}^n_4$. By the first part of Theorem 3.6 there is a homeomorphism $\tilde{\theta}\colon M'' \to M''$ so that $\bar{f}\tilde{\theta}$ is ε''-close to $\theta\bar{f}$, where the size of ε'' depends on the size of δ''. Define $\hat{f} = \bar{f}\tilde{\theta}^{-1}\colon M'' \to R^n$. This is a μ-fibration over $B^n_{1.5}$. Moreover we observe that there is a homotopy of $\hat{f}$ to $\bar{f}$, where the size of this homotopy depends on the size of ε''.

Define $\tilde{f}\colon M \to R^n$ to be $\hat{f}$ over $B^n_{1.5}$, to be f over $R^n - \mathring{B}^n_2$, and over $B^n_2 - \mathring{B}^n_{1.5}$ we let $\tilde{f}$ be defined by the homotopy $\tilde{f} \simeq \bar{f}$ described above. It is clear that $\tilde{f}$ fulfills our requirements. □

Before giving the statement of Theorem 5.2 we will have to introduce some more notation. Let X be a compact ANR and let $c(X)$ denote the *cone* over X. More precisely let $c(X) = X \times [0, \infty]/\sim$, where $\sim$ is the equivalence relation generated by $(x, 0) \sim (x', 0)$, for all $x, x' \in X$. Similarly let $\mathring{c}(X) = X \times [0, \infty)/\sim$ denote the *open cone* over X, and for any $t \in [0, \infty]$ let $c_t(X) = X \times [0, t]/\sim$ and $\mathring{c}_t(X) = X \times [0, t)/\sim$. In what follows X is a compact polyhedron.

THEOREM 5.2. *If* $\varepsilon > 0$, $m \geq 5$ *and* $n \geq 0$ *are given, then there exists a* $\delta > 0$ *such that the following is true: for every* $\mu > 0$ *there exists a* $\nu > 0$ *so that if* M^m *is a manifold and* $f\colon M \to \mathring{c}(X) \times R^n$ *is a* δ*-fibration over* $c_3(X) \times B^n_3$, *and a* ν*-fibration over* $[c_3(X) - \mathring{c}_{1/3}(X)] \times B^n_3$, *then there is a map* $\tilde{f}\colon M \to \mathring{c}(X) \times R^n$ *which is a* μ*-fibration over* $c_1(X) \times B^n_1$ *and which is* ε*-homotopic to* f rel $M - f^{-1}(\mathring{c}_{2/3}(X) \times \mathring{B}^n_3)$.

REMARKS ON PROOF. For $n \geq 1$ the proof is similar to the proof of Lemma 5.1. There are only two places in which any significant changes must be made. The first occurs in the wrapping up procedure. In the more general case that we are dealing with, Theorem 4.3 must be used in lieu of Lemma 4.1 in order to accomodate the $\mathring{c}(X)$-factor. The second occurs in the construction of the map $\bar{f} = \gamma f''$, where γ is a "radial squeeze" towards the origin. In this more general case, $\gamma\colon \mathring{c}(X) \times R^n \to \mathring{c}(X) \times R^n$ is a homeomorphism which is a composition $\gamma = \gamma_2\gamma_1$, where γ_1 gives a squeeze to R^n towards the origin and γ_2 gives a squeeze to $\mathring{c}(X)$ towards the vertex. This device of squeezing towards the vertex is amply illustrated in the case $n = 0$ below.

We now treat the case $n = 0$, which is not covered in the proof of Lemma 5.1. We are given $\varepsilon > 0$, $\mu > 0$ and a proper map $f\colon M \to \mathring{c}(X)$ which is a δ-fibration over $c_3(X)$ and a ν-fibration over $c_3(X) - \mathring{c}_{1/3}(X)$. Choose t close to 0 and let $\theta\colon \mathring{c}(X) \to \mathring{c}(X)$ be a homeomorphism which is supported on $c_{2/3}(X) - \mathring{c}_{t/2}(X)$ and which takes $c_t(X)$ to $c_{1/2}(X)$. This can be done so that all motion occurs along the $[0, \infty)$-direction in $\mathring{c}(X)$. By Lemma 3.5 there is a homeomorphism $\tilde{\theta}\colon M \to M$ for which $f\tilde{\theta}^{-1}$ is $\tilde{\delta}$-close to $\theta^{-1}f$, where the size of $\tilde{\delta}$ is dependent on the size of δ. Then $f' = \theta^{-1}f\tilde{\theta}\colon M \to \mathring{c}(X)$ is a $\tilde{\mu}$-fibration over $c_3(X) - \mathring{c}_t(X)$ and which is $\varepsilon/2$-homotopic to f rel $M - f^{-1}(\mathring{c}_{2/3}(X))$, where the size of $\tilde{\mu}$ depends on the size of μ. Let $\gamma\colon \mathring{c}(X) \to \mathring{c}(X)$ be a homeomorphism which is supported on $c_{2/3}(X)$ and which squeezes $c_t(X)$ close to the vertex. It is then clear that $\tilde{f} = \gamma f'$ fulfills our requirements. □

6. PROOF OF THEOREM 1 AND ITS COROLLARY

In order to illustrate our method of proof we first consider the special case in which B is a polyhedron and $\partial M = \emptyset$. We are given a β-fibration $f: M \to B$. We will construct sequences $\{f_i\}$ and $\{\beta_i\}$ such that $f_i: M \to B$ is a β_i-fibration, $\lim_{i\to\infty} \text{mesh}(\beta_i) = 0$, where

$$\text{mesh}(\beta_i) = \text{lub}\{\text{diam}(U) \mid U \in \beta_i\},$$

f_i is close to f, and $p = \lim_{i\to\infty} f_i: M \to B$ is a proper map. Certainly p must be close to f, and to see that p is an approximate fibration it suffices (by Proposition 2.2) to show that for every compactum $C \subset B$ and open cover γ of B, p is a γ-fibration over C. So suppose that we have maps $F: Z \times I \to C$ and $\tilde{F}_o: Z \to p^{-1}(C)$ such that $p\tilde{F}_o = F_o$. If i is large, then there is a small homotopy of $p|p^{-1}(C)$ to $f_i|p^{-1}(C)$. Since $p\tilde{F}_o = F_o$ this gives us a small homotopy of F_o to $f_i\tilde{F}_o$. By the homotopy extension theorem[†] we can get a small homotopy $F \simeq F'$ so that $F'_o = f_i\tilde{F}_o$. Since f_i has the β_i-lifting property we have a map $G: Z \times I \to M$ so that $G_o = \tilde{F}_o$ and f_iG is β_i-close to F'. Then pG is close to F as desired.

All that remains to be done for this case is the verification of the existence of the sequences $\{f_i\}$ and $\{\beta_i\}$. They are inductively chosen by the following

ASSERTION. *For every open cover* α *of* B *and* $m \geq 5$ *there exists an open cover* β *of* B *so that the following is true: if* M^m *is a manifold,* $\partial M = \emptyset$, *and* $f: M \to B$ *is a* β*-fibration, then for every open cover* γ *of* B *there exists a* γ*-fibration* $f': M \to B$ *which is* α*-close to* f.

[†] This is the usual homotopy extension theorem with an estimation built into it. See [7, Proposition 2.1] for a precise statement.

Proof. We will need a "handle decomposition" of B. Assume that B has a fixed triangulation and for each barycenter $b \in B$ let C_b be the closed star of b in the second barycentric subdivision of B. Recalling the cone notation of Section 5 we can regard each C_b as $c_1(X_b) \times B_1^k$, where X_b is a compact polyhedron and B_1^k is a neighborhood of b in the simplex of which b is the barycenter. By enlarging C_b slightly we obtain an open set in B of the form $\mathring{c}(X_b) \times R^k$ so that C_b is identified with $c_1(X_b) \times B_1^k \subset \mathring{c}(X_b) \times R^k$. (If b is the barycenter of a principal simplex, then we let $\mathring{c}(X_b) = \{b\}$ for convenience.) If the choices are correctly made, then for any fixed integer k the sets $\mathring{c}(X_b) \times R^k$ are pairwise disjoint.

If B is compact then the procedure is standard. Let $\dim B = n$ and consider the collection

$$\{\mathring{c}(X_b) \times R^n \mid b \text{ is the barycenter of some n-simplex}\}.$$

The open sets of this collection are pairwise disjoint, so the handle lemma of Section 5 can be applied to each of them (independently). Since $\mathring{c}(X_b) = \{b\}$ we only need Lemma 5.1. If β is sufficiently fine, then Lemma 5.1 yields a β_1-fibration $f_1\colon M \to B$ which is close to f and which is a γ_1-fibration over a neighborhood of

$$S_n = \cup\{C_b \mid b \text{ is the barycenter of some n-simplex}\},$$

where $\text{mesh}(\beta_1)$ depends on $\text{mesh}(\beta)$ and $\text{mesh}(\gamma_1)$ can be chosen as small as we want. This completes the first step of the construction.

The next step is to similarly treat the open sets $\mathring{c}(X_b) \times R^{n-1}$, where b is the barycenter of some (n-1)-simplex. In this case we use the full strength of the handle lemma (i.e., Theorem 5.2) to similarly find a β_2-fibration $f_2\colon M \to B$ which is close to f_1 and which is a γ_2-fibration over a neighborhood of $S_n \cup S_{n-1}$, where

$$S_{n-1} = \cup\{C_b \mid b \text{ is the barycenter of some (n-1)-simplex}\},$$

where $\text{mesh}(\beta_2)$ depends on $\text{mesh}(\beta_1)$, and where $\text{mesh}(\gamma_2)$ depends on $\text{mesh}(\gamma_1)$. This completes the second step of the construction. It is now clear that one can continue in this manner until all of the handles in B are exhausted.

If B is not compact then one writes B as a union of compact subpolyhedra, $B = \cup B_i$, so that $B_i \cup B_j = \emptyset$ for $|i - j| \geq 2$. Then the above procedure for the compact case is applied to the B_{2i-1} (independently) to produce a β_o-fibration $f_o: M \to B$ which is close to f and which is a γ_o-fibration over a neighborhood of $\cup B_{2i-1}$, where on any B_{2i-1} $\text{mesh}(\beta_o)$ depends on $\text{mesh}(\beta)$ and $\text{mesh}(\gamma_o)$ can be made as small as we want. Again applying the procedure for the compact case to the B_{2i} (independently) we produce our desired γ-fibration $f': M \to B$ which is α-close to f. This completes the proof of the special case in which B is a polyhedron and $\partial M = \emptyset$.

We now turn our attention to the case in which B is locally polyhedral and $\partial M = \emptyset$. If B is compact, then we can write $B = \cup_{i=1}^n B_i$, where B_i is a polyhedron which is open in B. Choose a compactum $C_i \subset B_i$ so that $\cup_{i=1}^n C_i = B$. If β is sufficiently fine, then by the proof of the special case above we can find open covers $\beta = \beta_1, \beta_2, \ldots, \beta_n = \gamma$ of B and maps $f_i: M \to B$ which are α-close to f and for which

(1) f_i is a β_i-fibration over a neighborhood of $C_1 \cup \cdots \cup C_i$,

(2) f_i is a β_i-fibration over a neighborhood of C_j, for $i + 1 \leq j \leq n$.

Thus $f_n: M \to B$ is a γ-fibration, and as in the proof of the special case above this implies that f is α-close to an approximate fibration as desired. If B is non-compact the procedure is similar. This completes the proof of Theorem 1. $\square$

Proof of Corollary. Suppose first that there is a compact manifold N^m, $\partial N = \emptyset$, and a homotopy equivalence $f': N \to K \times S^1$. Let $\nu: S^1 \to S^1$ be the map $\nu(z) = z^n$ (the standard n-fold covering map), and form the pull-back diagram

$$\begin{array}{ccc} M & \xrightarrow{f} & K \times S^1 \\ \downarrow & & \downarrow \mathrm{id} \times \nu \\ N & \xrightarrow{f'} & K \times S^1 \end{array} .$$

Then M^m is a compact manifold and $f\colon M \to K \times S^1$ is still a homotopy equivalence. More importantly, $\mathrm{proj} \circ f\colon M \to S^1$ is an ε-fibration, where $\varepsilon = \varepsilon(n)$ is small if n is large. This is an easy exercise using the basic properties of covering spaces. By Theorem 1 we have $\mathrm{proj} \circ f$ homotopic to an approximate fibration $p\colon M \to S^1$. The homotopy fiber of p is homotopy equivalent to the homotopy fiber of $\mathrm{proj}\colon K \times S^1 \to S^1$ because the homotopy fiber is an invariant of homotopy type. But the homotopy fiber of $\mathrm{proj}\colon K \times S^1 \to S^1$ is K.[†]

Next assume that K is homotopy equivalent to the fiber of some approximate fibration $p\colon M^m \to S^1$. Form the pullback diagram,

$$\begin{array}{ccc} \tilde{M} & \xrightarrow{\tilde{p}} & R \\ \downarrow & & \downarrow e \\ M & \xrightarrow{p} & S^1 \end{array}$$

Then $\tilde{p}$ is also an approximate fibration and $\tilde{M}$ is homotopy equivalent to K. In fact, there is a homotopy equivalence $h\colon \tilde{M} \to K \times R$ for which $\mathrm{proj} \circ h = \tilde{p}$. Thus h is a $p_R^{-1}(\varepsilon)$-equivalence, for all $\varepsilon > 0$, where $p_R = \mathrm{proj}\colon K \times R \to R$. By Theorem 4.2 we can wrap up to get a manifold N^m, $\partial N = \emptyset$, and a homotopy equivalence $N \simeq K \times S^1$. $\square$

CONCLUDING REMARKS. In the proof of Theorem 1 given above we have assumed that $\partial M = \emptyset$. In Section 1 we mentioned two variations in which $\partial M \neq \emptyset$. In the first variation we assume that f is already an approximate fibration over an open set in B which contains $f(\partial M)$. This gives us a buffer zone containing ∂M, and in the above proof we can choose the f_i so that $f_i = f$ over $f(\partial M)$. There is no change in the argument. In the second variation we assume that

---

[†]This is clarified by the fact that the homotopy fiber of p is shape equivalent to $p^{-1}(z)$, for all $z \in S^1$ [13].

$f|\partial M: \partial M \to B$ is the projection map of a locally trivial bundle or the projection map of a block bundle over simplices of arbitrarily small diameter. In both cases $f|\partial M: \partial M \to B$ is an approximate fibration and the bundle structure enables us to find engulfing homeomorphisms so that a version of Theorem 5.2 with ∂ can be done. Then in the above proof of the unbounded case we can choose the f_i so that $f_i = f$ on ∂M. The details of the locally trivial case bundle case are fairly straightforward, and for a further explanation of the block bundle case we refer the reader to the proof of the Handle Lemma in Appendix 3.

7. SOME LEMMAS FOR THEOREM 2

The purpose of this section is to establish some technical lemmas which will be needed in the proof of Theorem 2. As in Theorem 4.2 we will be concerned with wrapping up $p^{-1}(\delta)$-equivalences around S^1. In Lemma 7.1 below we have stated a result which is a special case of Theorem 4.2. The remainder of this section will be concerned with sharpening the conclusion of this result. Unfortunately it will be necessary to dredge up some of the details of the proof of Theorem 4.2.

For notation B will be a finite-dimensional polyhedron (but perhaps non-compact) and p will denote projection to $B \times R$. Also q will denote projection to $B \times S^1$. Assume that B has a fixed triangulation and let B have the metric derived from this triangulation (as in Lemma 3.4). Also let $B \times R$ have the metric which is the product of the metric on B with the standard metric on R. Here is our special case of Theorem 4.2.

LEMMA 7.1. *For every integer* $m \geq 5$ *and* $\varepsilon > 0$ *there exists a* $\delta > 0$ *so that if* M^m *is a manifold,* $\partial M = \emptyset$, F *is a compact* ANR, *and* $f: M \to B \times F \times R$ *is a* $p^{-1}(\delta)$*-equivalence over* $B \times F \times [-4, 4]$, *then there is a manifold* $\tilde{M}$, $\partial\tilde{M} = \emptyset$, *a* $q^{-1}(\varepsilon)$*-equivalence* $\tilde{f}: \tilde{M} \to B \times F \times S^1$, *and an open embedding* $\phi: f^{-1}(B \times F \times (-1, 1)) \to \tilde{M}$ *for which the following diagram commutes:*

$$\begin{array}{ccc} \tilde{M} & \xrightarrow{\quad \tilde{f} \quad} & B \times F \times S^1 \\ \Big\uparrow{\scriptstyle \phi} & & \Big\uparrow{\scriptstyle \mathrm{id} \times e} \\ f^{-1}(B \times F \times (-1, 1)) & \xrightarrow{\quad f \quad} & B \times F \times (-1, 1) \end{array}$$

As we have already indicated above the proof of this result is just the proof of Theorem 4.2, and that proof comes from Lemma 4.1. So our first task is to write out some notation which describes how $\tilde{f}: \tilde{M} \to B \times F \times S^1$ comes about. The basis of the definition of $\tilde{M}$ is a homeomorphism $h: M \to M$ which is supported on $f^{-1}(B \times F \times [-2.5, 2.5])$ and for which $p_R hf^{-1}(B \times F \times \{-2\})$ is close to 2. Then $\tilde{M} = Y/\sim$, where

$$Y = hf^{-1}(B \times F \times (-\infty, -2]) - f^{-1}(B \times F \times (-\infty, -2))$$

and $\sim$ is the equivalence relation generated by $x \sim h(x)$, for $x \in f^{-1}(B \times F \times \{-2\})$. This is precisely the definition given in Section 4.

The basis of the definition of $\tilde{f}$ is an isotopy h_t: id $\simeq h$ which is supported on $f^{-1}(B \times F \times [-2.5, 2.5])$. Let $\mu: Y \to [0, 1]$ be a Urysohn function which is 0 on $f^{-1}(B \times F \times \{-2\})$ and 1 on $hf^{-1}(B \times F \times \{-2\})$. Define $g: Y \to B \times F \times [-2, 2]$ so that the R-component of g, $p_R g$, satisfies (1) $p_R g$ is close to $p_R f|Y$, (2) $p_R f^{-1}(B \times F \times \{-2\}) = -2$, and (3) $p_R gf^{-1}(B \times F \times \{-2\}) = 2$. Define the $(B{\times}F)$-component of g, $p_1 g$, by $p_1 g(x) = p_1 fh^{-1}_{\mu(x)}(x)$, where p_1 is projection to $B \times F$. Then g factors through the appropriate equivalence relations to yield our $\tilde{f}: \tilde{M} \to B \times F \times S^1$. It is clear that the homotopy class of $\tilde{f}$ is independent of the choice of the Urysohn function μ. It is our intention to show that the homotopy class of $\tilde{f}$ is independent of all choices.

LEMMA 7.2. $\tilde{f}$ *is well-defined.*

Proof. We want to show that if $\bar{h}: M \to M$ is another homeomorphism which is supported on $f^{-1}(B \times F \times [-2.5, 2.5])$ and for which $p_R hf^{-1}(B \times F \times \{-2\})$ is close to 2, and $\bar{h}_t$: id $\simeq \bar{h}$ is another isotopy which is supported on $f^{-1}(B \times F \times [-2.5, 2.5])$, then there is a homeomorphism $k: \tilde{M} \to \bar{M}$ for which $\bar{f}k \simeq \tilde{f}$. Here $\bar{M} = \bar{Y}/\sim$ is defined in analogy with $\tilde{M} = Y/\sim$, and $\bar{f}: \bar{M} \to B \times F \times S^1$ is defined in analogy with $\tilde{f}$.

By Lemma 3.5 there is a homeomorphism $\alpha: M \to M$ for which

$$f^{-1}(B \times F \times [-2.5, \infty)) \subset \alpha f^{-1}(B \times F \times [1.8, \infty))$$

and for which there is an isotopy α_t: id $\simeq \alpha$ which is the identity on

$$f^{-1}(B \times F \times (-\infty, -2.6]) \cup f^{-1}(B \times F \times [1.9, \infty)).$$

Then $\alpha^{-1}\bar{h}h^{-1}\alpha: M \to M$ is a homeomorphism which is supported on $f^{-1}(B \times F \times [1.8, 2.5])$ and which restricts to a homeomorphism $j: Y \to \bar{Y}$ which is id on $f^{-1}(B \times F \times \{-2\})$ and $\bar{h}h^{-1}$ on $hf^{-1}(B \times F \times \{-2\})$. It will suffice to prove that $p_1\bar{g}j \simeq p_1 g$ rel E, where

$$E = f^{-1}(B \times F \times \{-2\}) \cup hf^{-1}(B \times F \times \{-2\}).$$

We have a homotopy $m_t: p_1\bar{g}j \simeq p_1 fj$ defined by homotoping the Urysohn function $\bar{\mu}$ on $\bar{Y}$ to 0. Note that $m_t = p_1 f$ on $f^{-1}(B \times F \times \{-2\})$ and $m_t = p_1 f\bar{h}_{1-t}^{-1}\bar{h}h^{-1}$ on $hf^{-1}(B \times F \times \{-2\})$. We have a homotopy $n_t: p_1 fj \simeq p_1 f|Y$ defined by $n_t(x) = p_1 f\alpha^{-1}\bar{h}_{1-t}h_{1-t}^{-1}\alpha(x)$. Note that $n_t = p_1 f$ on $f^{-1}(B \times F \times \{-2\})$ and $n_t = p_1 f\alpha^{-1}\bar{h}_{1-t}h_{1-t}^{-1}$ on $hf^{-1}(B \times F \times \{-2\})$. Form the "juxtaposition" of m_t and n_t, i.e.,

$$m_t * n_t = \begin{cases} m_{2t}, & 0 \leq t \leq 1/2 \\ n_{2t-1}, & 1/2 \leq t \leq 1 . \end{cases}$$

Then $m_t * n_t: Y \to B \times F$ is a homotopy from $p_1\bar{g}j$ to $p_1 f|Y$ which is constantly $p_1 f$ on $f^{-1}(B \times F \times \{-2\})$. We also have a homotopy $q_t: p_1 g \simeq p_1 f|Y$ defined by homotoping the Urysohn function μ on Y to 0. Note that $q_t = p_1 f$ on $f^{-1}(B \times F \times \{-2\})$ and $q_t = p_1 fh_{1-t}^{-1}$ on $hf^{-1}(B \times F \times \{-2\})$. Thus $(m_t * n_t) * q_{1-t}: Y \to B \times F$ is a homotopy from $p_1\bar{g}j$ to $p_1 g$ which is constantly $p_1 f$ on $f^{-1}(B \times F \times \{-2\})$. On $hf^{-1}(B \times F \times \{-2\})$ this homotopy is $p_1 f\bar{h}_{1-t}^{-1}\bar{h}h^{-1} * p_1 f\alpha^{-1}\bar{h}_{1-t}h_{1-t}^{-1} * p_1 fh_t^{-1}$. It is a homotopy from $p_1 fh^{-1}$ to $p_1 fh^{-1}$. If it is homotopic rel $\{0, 1\}$ through a 1-parameter family of homotopies to the constant homotopy $p_1 fh^{-1} \simeq p_1 fh^{-1}$, then the Homotopy Extension Theorem will imply that the homotopy $(m_t * n_t) * q_{1-t}$ can be homotoped to a new homotopy $p_1\bar{g}j \simeq p_1 g$ rel E as desired.

All we need is a homotopy of $\bar{h}_{1-t}^{-1}\bar{h}h^{-1} * \alpha^{-1}\bar{h}_{1-t}h_{1-t}^{-1} * h_t^{-1}$ (on $hf^{-1}(B \times F \times \{-2\})$ to the constant homotopy $h^{-1} \simeq h^{-1}$ rel $\{0, 1\}$. Here is the sequence of steps:

$$\bar{h}_{1-t}^{-1}\bar{h}h^{-1} * \alpha^{-1}\bar{h}_{1-t}h_{1-t}^{-1} * h_t^{-1} \simeq \bar{h}_{1-t}^{-1}\bar{h}h^{-1} * \bar{h}_{1-t}h_{1-t}^{-1} * h_t^{-1}$$

$$\simeq h^{-1} * (h_{1-t}^{-1} * h_t^{-1})$$

$$\simeq h^{-1} * h^{-1} \simeq h^{-1} \quad \text{(the constant homotopy).}$$

Details are left to the reader. □

REMARKS. 1. In the definition of $\tilde{f}: \tilde{M} \to B \times F \times S^1$ we may assume that $p_B fh_t: M \to B$ is a γ-homotopy, for γ small. If we additionally assume that $p_B f\bar{h}_t: M \to B$ is a γ-homotopy in the above proof, then the homotopy $\bar{f}k \simeq \tilde{f}$ which was constructed is a $(p_B\tilde{f})^{-1}(\gamma')$-homotopy, for γ' small.

2. In the above construction there is nothing sacred about the numbers -2 and 2. We could perform this construction for numbers $-r_1$ and r_2, $-3 \leq -r_1 < 0 < r_2 \leq 3$, and thereby obtain $\tilde{f}: \tilde{M} \to B \times F \times S^1$. By an argument similar to Lemma 7.2 above this is well-defined (i.e., independent of r_1 and r_2).

3. If we replace f by $f_1: M \to B \times F \times R$ which is $p^{-1}(\gamma)$-homotopic to f, for γ small, then we obtain $\tilde{f}_1: \tilde{M}_1 \to B \times F \times S^1$ in a manner analogous to $\tilde{f}: \tilde{M} \to B \times F \times S^1$. Just as in Lemma 7.2 we can find a homeomorphism $k: \tilde{M} \to \tilde{M}_1$ for which $\tilde{f}_1 k \simeq \tilde{f}$ via a $(p_B\tilde{f})^{-1}(\gamma')$-homotopy, for γ' small.

4. Finally we mention how all of the above can be deduced from the twist-gluing construction of [21, Section 5]. Given a $p^{-1}(\delta)$-equivalence over $B \times F \times [-4,4]$, $f: M \to B \times F \times R$, it is shown in [7 , Section 5] how to build an open set U,

$$f^{-1}(B \times F \times [-1, 1]) \subset U \subset f^{-1}(B \times F \times [-2, 2]),$$

so that there is a proper homotopy equivalence $g: U \to B \times F \times R$ which agrees with f on $f^{-1}(B \times F \times [-1, 1])$. Then the twist-gluing construction of [21] (twist = identity) yields $\bar{f}: \bar{M} \to B \times F \times S^1$. It is not hard to see that this is equivalent to our $\tilde{f}: \tilde{M} \to B \times F \times S^1$.

We will also need the following result concerning $\tilde{f}: \tilde{M} \to B \times F \times S^1$.

LEMMA 7.3. $\tilde{f}\colon \tilde{M} \to B \times F \times S^1$ *is invariant under passage to finite covers.*

Proof. We first explain just what this means. Let $\nu\colon S^1 \to S^1$ be a standard finite cover and form the pull-back diagram,

$$\begin{array}{ccc} \hat{M} & \xrightarrow{\hat{f}} & B \times F \times S^1 \\ \downarrow & & \downarrow \mathrm{id} \times \nu \\ \tilde{M} & \xrightarrow{\tilde{f}} & B \times F \times S^1 \end{array} .$$

We want a homeomorphism $k\colon \tilde{M} \to \hat{M}$ for which $\hat{f}k \simeq \tilde{f}$. It will suffice to treat the case in which ν is a two-fold cover.

Using the notation which was introduced for the definition of $\tilde{f}$ let $\hat{Y} = Y \cup h(Y)$. Then $\hat{M}$ is homeomorphic to $\hat{Y}/\sim$, where $\sim$ is the equivalence relation generated by $x \sim h^2(x)$, for $x \in f^{-1}(B \times F \times \{-2\})$. Define $\hat{g}\colon \hat{Y} \to B \times F \times [-2, 2]$ by

$$p_R\hat{g} = \begin{cases} \frac{1}{2}\,(p_R g - 2), & \text{on } Y \\ \frac{1}{2}\,(p_R g h^{-1} + 2), & \text{on } h(Y) \end{cases}$$

and

$$p_1\hat{g} = \begin{cases} p_1 g \quad , & \text{on } Y \\ p_1 g h^{-1} \quad , & \text{on } h(Y) \end{cases}$$

Then $\hat{g}$ factors through the appropriate equivalence relations to define $\hat{f}\colon \hat{M} \to B \times F \times S^1$.

It is clear that there is a Urysohn function μ on $Y \cup h(Y)$ which is 0 on $f^{-1}(B \times F \times \{-2\})$, 1 on $h^2 f^{-1}(B \times F \times \{-2\})$, and for which $p_1\hat{g}(x) = p_1 f k^{-1}_{\mu(x)}(x)$, where k_t^{-1} is the juxtaposition $h_t^{-1} * h^{-1}h_t^{-1}$. If we put $\bar{h}_t = h_t * h_t h$, $\bar{h} = h^2$, and $\bar{Y} = \hat{Y}$, then repeating the proof of Lemma 7.2 we get a homeomorphism $j\colon Y \to \bar{Y}$ which induces $k\colon \tilde{M} \to \hat{M}$. □

Finally we will treat a realtive version of Lemma 7.1 in which we are dealing with manifolds that have boundary. Again B will be a finite dimensional polyhedron.

LEMMA 7.4. *For every integer* $m \geq 5$ *and* $\varepsilon > 0$ *there exists a* $\delta > 0$ *such that if* M^m *is a manifold,* (F, F_o) *is a compact* ANR *pair, and* $f: M \to B \times F \times R$ *is a* $p^{-1}(\delta)$*-equivalence over* $B \times F \times [-4, 4]$ *for which* $f|\partial M: \partial M \to B \times F_o \times R$ *is a homeomorphism, then there exists a manifold* $\tilde{M}$, *a* $q^{-1}(\varepsilon)$*-equivalence* $\tilde{f}: \tilde{M} \to B \times F \times S^1$ *such that* $\tilde{f}|\partial\tilde{M}: \partial\tilde{M} \to B \times F_o \times S^1$ *is a homeomorphism, and an open embedding* $\phi: f^{-1}(B \times F \times (-1, 1)) \to \tilde{M}$ *for which the following diagram commutes:*

$$\begin{array}{ccc} \tilde{M} & \xrightarrow{\tilde{f}} & B \times F \times S^1 \\ \phi \uparrow & & \uparrow \mathrm{id} \times e \\ f^{-1}(B \times F \times (-1, 1)) & \xrightarrow{f} & B \times F \times (-1, 1) \end{array}$$

REMARKS ON PROOF. Tracing through the proof of Lemma 7.1 we see that the proof of Lemma 7.4 follows mechanically from the proof of Lemma 7.1 provided that we have a relative version of the engulfing isotopies of the type encountered in Lemma 3.4. For this we refer the reader to the Concluding Remarks at the end of Section 3. □

CONCLUDING REMARKS. As might be expected the proofs of Lemmas 7.2 and 7.3 easily yield the corresponding relative versions for M with boundary.

8. PROOF OF THEOREM 2

Let F^m be a compact manifold. In what follows we will first treat the case in which $\partial F = \emptyset$, and then at the end of the proof we will indicate what changes have to be made to accomodate the bounded case. Fortunately these changes are minor and routine. So our first task will be to start with a $p^{-1}(\delta)$-equivalence $f: M^{m+n} \to R^n \times F^m$, δ small and $\partial M = \emptyset = \partial F$, and show how this gives rise to an element $\sigma(f) \in \mathcal{S}(T^n \times F)$. Then we will prove that $\sigma(f)$ fulfills our requirements. In what follows p will denote projection to R^n and q will denote projection to T^n.

I. DEFINITION OF $\sigma(f)$. Applying the wrapping up procedure of Section 7 n times we can wrap up f to get a $q^{-1}(\tilde{\delta})$-equivalence $\tilde{f}: \tilde{M} \to T^n \times F$. Note that $\tilde{M}$ is a compact manifold and we may regard $f^{-1}(\mathring{B}^n_{1.8} \times F)$ as an open subset of $\tilde{M}$ so that the following diagram commutes:

$$\begin{array}{ccc} \tilde{M} & \xrightarrow{\tilde{f}} & T^n \times F \\ \uparrow & & \uparrow {\scriptstyle e^n \times id} \\ f^{-1}(\mathring{B}^n_{1.8} \times F) & \xrightarrow{f} & \mathring{B}^n_{1.8} \times F \end{array}$$

Then we define $\sigma(f) = [\tilde{f}] \in \mathcal{S}(T^n \times F)$.

II. IF $\sigma(f)$ VANISHES, THEN f IS $p^{-1}(\varepsilon)$-HOMOTOPIC TO A HOMEOMORPHISM. We are given that the $q^{-1}(\tilde{\delta})$-equivalence $f: M \to T^n \times F$ is homotopic to a homeomorphism $h: \tilde{M} \to T^n \times F$. Form the pullback diagram

$$\begin{array}{ccc} \hat{M} & \xrightarrow{\hat{f}} & R^n \times F \\ \downarrow & & \downarrow {\scriptstyle e^n \times id} \\ \tilde{M} & \xrightarrow{\tilde{f}} & T^n \times F \end{array} .$$

The homotopy $\tilde{f} \simeq h$ lifts to a bounded homotopy[†] $\hat{f} \simeq \hat{h}$, where $\hat{h}: \hat{M} \to R^n \times F$ is a homomorphism covering h. By Proposition 2.1 $\hat{f}$ must be a $p^{-1}(\gamma)$-equivalence, for γ small and dependent on $\tilde{\delta}$. Combining the above diagram with the diagram from step I we can choose notation so that $\hat{f} = f$ over $\mathring{B}^n_2 \times F$.

Choose $K > 0$ large and let $f_1: \hat{M} \to R^n \times F$ be defined so that (i) it agrees with $\hat{f}$ over $B^n_{K-1} \times F$, (ii) it agrees with $\hat{h}$ over $(R^n - \mathring{B}^n_K) \times F$, and (iii) over $(B^n_K - \mathring{B}^n_{K-1}) \times F$ it is defined by using the bounded homotopy $\hat{h} \simeq \hat{f}$. (For this it is necessary to assume that $\hat{h}^{-1}((R^n - \mathring{B}^n_K) \times F)$ and $\hat{f}^{-1}(B^n_{K-1} \times F)$ are disjoint.) We now refer the reader to the last half of the proof of Lemma 5.1. Specifically we refer to the part which follows the proof of the Assertion. Repeating this construction it is easy to get an open embedding $g: f^{-1}(\mathring{B}^n_{1.8} \times F) \to R^n \times F$ which is $p^{-1}(\varepsilon_1)$-homotopic to $f| f^{-1}(\mathring{B}^n_{1.8} \times F)$, for some small $\varepsilon_1 > 0$. Then we obtain a $p^{-1}(\bar{\varepsilon})$-equivalence $\bar{f}: M \to R^n \times F$ which is a homomorphism over $\mathring{B}^n_{1.5} \times F$ and which is $p^{-1}(\bar{\varepsilon})$-homotopic to f, for $\bar{\varepsilon}$ small. To finish this step it will suffice to prove that $\bar{f}$ is $p^{-1}(\gamma)$-homotopic to a homomorphism, for γ small.

Since $\bar{f}$ is a $p^{-1}(\bar{\varepsilon})$-equivalence, $p\bar{f}: M \to R^n$ is easily seen to be a $2\bar{\varepsilon}$-fibration. Also $p\bar{f}$ is an α-fibration over $\mathring{B}^n_{1.5}$, for every open cover α of $\mathring{B}^n_{1.5}$. It follows from Theorem 1 that if $\bar{\varepsilon}$ is sufficiently small, then there is a small homotopy of $p\bar{f}$ rel $\bar{f}^{-1}(B^n_{1.4} \times F)$ to an approximate fibration $p': M \to R^n$. The small homotopy $p\bar{f} \simeq p'$ lifts to a homotopy $\bar{f} \simeq f^1$ rel $\bar{f}^{-1}(B^n_{1.4} \times F)$. Thus f^1 is a homeomorphism over $B^n_{1.2} \times F$ and $pf^1 = p'$ is an approximate fibration. By Proposition 2.3, f^1 is a $p^{-1}(\alpha)$-equivalence, for every open cover α of R^n. It will suffice to prove that f^1 is $p^{-1}(\gamma)$-homotopic to a homeomorphism, for γ small.

Again using the ideas of the last half of the proof of Lemma 5.1 we can find homeomorphisms $\theta: R^n \to R^n$, $\tilde{\theta}: M \to M$ so that $f^2 = (\theta \times id)^{-1} f^1 \tilde{\theta}: M \to R^n \times F$ is a homeomorphism over $B^n_{2.2} \times F$ and so that $f^2 \simeq f^1$ rel $(f^1)^{-1}(B^n_1 \times F)$.

[†] A *bounded homotopy* into $R^n \times F$ is a $p^{-1}(c)$-homotopy, for some constant $c > 0$.

It should be observed that θ is nothing but a radially-defined contraction which takes $B^n_{2.2}$ to $B^n_{1.2}$, and which is id on $B^n_{1.1}$. Also the R^n-component of the homotopy $f^2 \simeq f^1$ is small. Note that f^2 is still a $p^{-1}(\alpha)$-equivalence, for every open cover α of R^n. Repeating this construction we get $f^2 \simeq f^3$ rel $(f^2)^{-1}(B^n_2 \times F)$ so that f^3 is a homeomorphism over $B^n_{3.2} \times F$, f^3 is a $p^{-1}(\alpha)$-equivalence for every α, and so that the R^n-component of this homotopy is small. It is then clear that we can continue this process to get our desired $p^{-1}(\gamma)$-homotopy of f^1 to the homeomorphism $\lim_{n\to\infty} f^n$.

III. $\sigma(f) \in S_0(T^n \times F)$. Choose any integer $k \geq 1$. We want to prove that $\sigma(f)$ is invariant under k-fold covers of T^n. Write $R^n = R_1 \times R^{n-1}$ and perform the first wrapping up of f to obtain $f_1\colon M_1 \to S^1_1 \times R^{n-1} \times F$. By Lemma 7.3 f_1 is invariant under k-fold covers of S^1_1 *with control in* R^{n-1}. This means that if $\mu\colon S^1_1 \to S^1_1$ is the standard k-fold cover and we form the pullback,

$$\begin{array}{ccc} \tilde{M}_1 & \xrightarrow{\tilde{f}_1} & S^1_1 \times R^{n-1} \times F \\ \downarrow & & \downarrow{\scriptstyle \mu \times \mathrm{id}} \\ M_1 & \xrightarrow{f_1} & S^1_1 \times R^{n-1} \times F \end{array},$$

then there is a homeomorphism $h\colon \tilde{M}_1 \to M_1$ for which $f_1 h \simeq \tilde{f}_1$ via a homotopy whose projection to R^{n-1} is small. This is the first step of the proof.

For the next step write $R^{n-1} = R_2 \times R^{n-2}$ and perform a wrapping up of f_1 to obtain $f_2{:}M_2$ $S^1_1 \times S^1_2 \times R^{n-2} \times F$. We will prove that f_2 is invariant under k-fold covers of $S^1_1 \times S^1_2$ with control in R^{n-2}. We already know that f_2 is invariant under k-fold covers of S^1_2 with control in $S^1_1 \times R^{n-2}$. We will show that it suffices to prove that f_2 is invariant under k-fold covers of S^1_1 with control in R^{n-2}. To do this let $\mu\colon S^1_i \to S^1_i$ be the standard k-fold cover and form the pullbacks,

$$\begin{array}{ccc} M'' & \xrightarrow{f''} & S_1^1 \times S_2^1 \times R^{n-2} \times F \\ \downarrow & & \downarrow \scriptstyle{id \times \mu \times id} \\ M' & \xrightarrow{f'} & S_1^1 \times S_2^1 \times R^{n-2} \times F \\ \downarrow & & \downarrow \scriptstyle{\mu \times id} \\ M_2 & \xrightarrow{f_2} & S_1^1 \times S_2^1 \times R^{n-2} \times F \end{array} \quad \text{and} \quad \begin{array}{ccc} \bar{M} & \xrightarrow{\bar{f}} & S_1^1 \times S_2^1 \times R^{n-2} \times F \\ \downarrow & & \downarrow \scriptstyle{id \times \mu \times id} \\ M_2 & \xrightarrow{f_2} & S_1^1 \times S_2^1 \times R^{n-2} \times F \end{array}$$

We want a homeomorphism $j: M'' \to M_2$ for which $f_2 j \simeq f''$ via a homotopy whose projection to R^{n-2} is small. We are given homeomorphisms $h: M' \to M_2$ and $k: \bar{M} \to M_2$ for which $f_2 h \simeq f'$ and $f_2 k \simeq \bar{f}$ via homotopies whose projection to R^{n-2} are small. Consider the pullback,

$$\begin{array}{ccc} \hat{M} & \xrightarrow{\hat{f}} & S_1^1 \times S_2^1 \times R^{n-2} \times F \\ \downarrow & & \downarrow \scriptstyle{id \times \mu \times id} \\ M' & \xrightarrow{f_2 h} & S_1^1 \times S_2^1 \times R^{n-2} \times F \end{array}$$

The homotopy $f_2 h \simeq f'$ (with control in R^{n-2}) yields a homeomorphism $u: M'' \to \hat{M}$ for which $\hat{f} u \simeq f''$ (with control in R^{n-2}). But the above pullback factors into the two pullbacks,

$$\begin{array}{ccccc} \hat{M} & \xrightarrow{\tilde{h}} & \bar{M} & \xrightarrow{\bar{f}} & S_1^1 \times S_2^1 \times R^{n-2} \times F \\ \downarrow & & \downarrow & & \downarrow \scriptstyle{id \times \mu \times id} \\ M' & \xrightarrow{h} & M_2 & \xrightarrow{f_2} & S_1^1 \times S_2^1 \times R^{n-2} \times F, \end{array}$$

where $\tilde{h}$ is a homeomorphism covering h. Thus $j = k\tilde{h}u: M'' \to M_2$ is our desired homeomorphism.

So all we have left to do is prove that f_2 is invariant under k-fold covers of S_1^1 with control in R^{n-2}. If $e: R_2 \to S_2^1$ is the usual covering projection form the pullback,

$$\begin{array}{ccc} \tilde{M}_2 & \xrightarrow{\tilde{f}_2} & S_1^1 \times R_2 \times R^{n-2} \times F \\ \downarrow & & \downarrow \mathrm{id} \times e \times \mathrm{id} \\ M_2 & \xrightarrow{f_2} & S_1^1 \times S_2^1 \times R^{n-2} \times F. \end{array}$$

By the commutativity of diagrams there is a homeomorphism

$$u: \tilde{f}_2^{-1}(S_1^1 \times (-1.8,\ 1.8) \times R^{n-2} \times F) \to f_1^{-1}(S_1^1 \times (-1.8,\ 1.8) \times R^{n-2} \times F)$$

for which $f_1 u = \tilde{f}_2$. By an argument similar to that used in step II we can "stretch out" u to obtain a homeomorphism $\tilde{u}: \tilde{M}_2 \to M_1$ for which $f_1\tilde{u} \simeq \tilde{f}_2$ via a homotopy whose projection to $S_1^1 \times R_2 \times R^{n-2}$ is small. Since f_1 is invariant under k-fold covers of S_1^1 with control in $R_2 \times R^{n-2}$ it follows that f_2 is also invariant under k-fold covers of S_1^1 with control in $R_2 \times R^{n-2}$. We will use this fact to prove that f_2 is invariant under k-fold covers of S_1^1 with control in R^{n-2}.

Let $\mu: S_1^1 \to S_1^1$ be the standard k-fold cover and form the pullback,

$$\begin{array}{ccc} \bar{M}_2 & \xrightarrow{\bar{f}_2} & S_1^1 \times R_2 \times R^{n-2} \times F \\ \downarrow & & \downarrow \mu \times \mathrm{id} \\ \tilde{M}_2 & \xrightarrow{\tilde{f}_2} & S_1^1 \times R_2 \times R^{n-2} \times F \ . \end{array}$$

By the preceding paragraph there is a homeomorphism $h: \bar{M}_2 \to \tilde{M}_2$ so that $\tilde{f}_2 h \simeq \bar{f}_2$ via a homotopy whose projection to $R_2 \times R^{n-2}$ is small. By taking the natural identification of $\tilde{f}_2^{-1}(S_1^1 \times \{-2\} \times R^{n-2} \times F)$ with $\tilde{f}_2^{-1}(S_1^1 \times \{2\} \times R^{n-2} \times F)$ we obtain M_2. By taking the natural identification of $\bar{f}_2^{-1}(S_1^1 \times \{-2\} \times R^{n-2} \times F)$ with $\bar{f}_2^{-1}(S_1^1 \times \{2\} \times R^{n-2} \times F)$ we obtain the k-fold cover of M_2,

$$\begin{array}{ccc} \hat{M}_2 & \xrightarrow{\hat{f}_2} & S_1^1 \times S_2 \times R^{n-2} \times F \\ \downarrow & & \downarrow \mu \times \mathrm{id} \\ M_2 & \xrightarrow{f_2} & S_1^1 \times S_2 \times R^{n-2} \times F. \end{array}$$

Since M_2 is formed from a wrapping up of M_1 this can be transferred to $\tilde{M}_2$ to obtain a homeomorphism $\tilde{\theta}: \tilde{M}_2 \to \tilde{M}_2$ which is supported on $\tilde{f}_2^{-1}(S_1^1 \times [-2.5, 2.5] \times R^{n-2} \times F)$ and which makes the natural identification of $\tilde{f}_2^{-1}(S_1^1 \times \{-2\} \times R^{n-2} \times F)$ with $\tilde{f}_2^{-1}(S_1^1 \times \{2\} \times R^{n-2} \times F)$. Then $\tilde{\theta}$ lifts to give a homeomorphism $\bar{\theta}: \bar{M} \to \bar{M}$ which is supported on $\bar{f}^{-1}(S_1^1 \times [-2.5, 2.5] \times R^{n-2} \times F)$ and which makes the natural identification of $\bar{f}^{-1}(S_1^1 \times \{-2\} \times R^{n-2} \times F)$ with $\bar{f}^{-1}(S_1^1 \times \{2\} \times R^{n-2} \times F)$. Thus $\bar{\theta}$ enables us to wrap up $\bar{M}_2$ to get $\hat{M}_2$. The homeomorphism $h^{-1}\tilde{\theta}h$ also enables us to wrap up $\bar{M}_2$ to get M_2, and by the uniqueness, Lemma 7.2, we must have a homeomorphism $g: \hat{M}_2 \to M_2$ for which $f_2 g \simeq \hat{f}_2$ via a homotopy whose projection to R^{n-2} is small. This completes the proof that f_2 is invariant under k-fold covers of $S_1^1 \times S_2^1$, and therefore the second step of our proof is done.

The reader should now be convinced that the above process can be inductively carried out until all the wrappings are completed. At any given stage we are faced with $f_r: M_r \to S_1^1 \times \cdots \times S_r^1 \times R^{n-r} \times F$ which is invariant under k-fold covers of $T^r = S_1^1 \times \cdots \times S_r^1$, with control in R^{n-r}. Performing the next wrapping up we get $f_{r+1}: M_{r+1} \to T^r \times S_{r+1}^1 \times R^{n-r-1} \times F$. Then we repeat (word-for-word) the above proof to show that f_{r+1} is invariant under k-fold covers of $T^r \times S_{r+1}^1$ with control in R^{n-r-1}. Thus the process can be continued until we reach $\tilde{f}: \tilde{M} \to T^n \times F$ which is invariant under k-fold covers of T^n. This completes the proof of Theorem 2 in the unbounded case.

For the bounded case we are given a $p^{-1}(\delta)$-equivalence $f: M \to R^n \times F$ for which $f|\partial M: \partial M \to R^n \times \partial F$ is a homeomorphism. By n-applications of Lemma 7.4 we can wrap up f to get a $q^{-1}(\tilde{\delta})$-equivalence $\tilde{f}: \tilde{M} \to T^n \times F$ which is a homeomorphism from $\partial\tilde{M}$ to $T^n \times \partial F$ and for which $f^{-1}(\mathring{B}^n_{1.8} \times F)$ is an open subset of $\tilde{M}$ making the following diagram commute:

$$\begin{array}{ccc} \tilde{M} & \xrightarrow{\ \tilde{f}\ } & T^n \times F \\ \uparrow & & \uparrow {\scriptstyle e^n \times id} \\ f^{-1}(\mathring{B}^n_{1.8} \times F) & \xrightarrow{\ f\ } & \mathring{B}^n_{1.8} \times F \end{array}$$

Then $\sigma(f) = [\tilde{f}] \in \mathcal{S}(T^n \times F)$. We leave it to the reader to check through steps II and III above to show that $\sigma(f) \in \mathcal{S}_o(T^n \times F)$ and to show that $\sigma(f) = 0$ implies that f is $p^{-1}(\varepsilon)$-homotopic to a homeomorphism rel ∂M. There are no significant changes in the argument. $\square$

9. PROOF OF THEOREM 3[†]

Let F^m be a compact simply connected manifold, possibly with boundary. We will determine $\mathcal{S}(T^n \times F)$, $m + n \geq 5$, and then obtain the computation of $\mathcal{S}_o(T^n \times F)$ as a special case. We adopt the language of the surgical appendices B and C of Essay V in [17]. Here is the main result.

PROPOSITION 9.1. *$\mathcal{S}(T^n \times F)$ is in 1-1 correspondence with*

$$[T^n \times F_o, T^n \times \partial F; G/TOP, *].$$

The correspondence comes from restricting the normal invariant to $T^n \times F_o$.

Proof. To make life simpler we will use $[M/\partial; G/TOP]$ to stand for

$$[M, \partial M; G/TOP, *].$$

Now consider the commutative diagram of abelian groups whose rows are Sullivan-Wall exact sequences and $\pi = \pi_1(T^n)$ [17, p.269 and 283]:

$$\begin{array}{ccccccccc}
\cdots \xrightarrow{\theta'} & L_{m+n+1}(\pi) & \xrightarrow{d} & \mathcal{S}(T^n \times F) & \xrightarrow{\tau} & [T^n \times F/\partial; G/TOP] & \xrightarrow{\theta} & L_{m+n}(\pi) \\
 & \uparrow = & & \uparrow & & j \uparrow \text{ injective} & & \uparrow = \\
\cdots \underset{\cong}{\xrightarrow{\theta'}} & L_{m+n+1}(\pi) & \xrightarrow{d} & \mathcal{S}(T^n \times B^m) & \xrightarrow{\tau} & [T^n \times B^m/\partial; G/TOP] & \underset{\cong}{\xrightarrow{\theta}} & L_{m+n}(\pi) \\
 & & & \parallel & & & & \\
 & & & 0 & & & &
\end{array}$$

Here B^m is an m-disc in int F^m and the vertical maps correspond to inclusion $T^n \times B^m \to T^n \times F$. In the bottom row θ and θ' are isomorphisms by [17, p.274 and 275]. It follows that, in the top row, τ maps $\mathcal{S}(T^n \times F)$ isomorphically onto

[†] With the exception of some minor notational changes and minor rewording, this section was written by L. C. Siebenmann.

$$\text{cokernel}(j) = \text{kernel of upper } \theta.$$

We can replace F_o by the compactum $F - \mathring{B}^m$ and do so henceforth. It remains to show that cokernel(j) is

$$[(T^n \times F_o)/\partial;\ G/TOP].$$

To check this we use the period four cohomology theory with coefficients G/TOP,

$$H^i(X, Y) = [\Sigma^{4s-i}(X/Y);\ G/TOP], \quad 4s \geq i,$$

(see [17, Appendix C]) defined here for compact connected pairs, and apply it to the triple $T^n \times (F, F_o, \partial F) = (Y, Y_o, \partial Y)$ getting the exact sequence

$$H^i(Y, Y_o) \xrightarrow{j'} H^i(Y, \partial Y) \to H^i(Y_o, \partial Y) \overset{\delta}{\to} H^{i+1}(Y, Y_o) \xrightarrow{j''} H^{i+1}(Y, \partial Y) \to \cdots.$$

Note that for $i \equiv 0 \bmod 4$, j' is the map j above. We wish to calculate (for $i \equiv 0 \bmod 4$) that cokernel(j') is $H^i(Y_o, \partial Y)$. This is so (for all i) because the map j'' coincides with the injective map j in a surgery exact sequence diagram similar to the one above.

Indeed, with $i = 4s$ for example, it is the surgery diagram above with $I^{4s-1} \times F$ in place of F. □

GENERALIZATIONS. The above proposition and its proof hold good if T^n is any compact manifold, possibly with boundary, such that $\mathcal{S}(T^n \times B^m) = 0$ for $m + n \geq 5$. This is the case, for example, if T^n is a $K(G, 1)$ with G poly-Z, or a π_1 of a compact surface, or a π_1 of a fibered knot, see [17, p.284-287].

One can go further by formulating the (generalized) proposition and its proof modulo a Serre class of abelian groups, cf. [17, p.286].

It is not known to what extent the proposition holds when T^n is a closed manifold of constant negative curvatures $(n > 2)$.

COROLLARY. $\mathcal{S}_o(T^n \times F^m)$ *is in* 1-1 *correspondence with* $[F_o/\partial;\ G/TOP]$.

Proof. It follows from [17, p.268] that

$$[(T^n \times F_o)/\partial;\ G/TOP] \approx \bigoplus_i \binom{n}{i} [\Sigma_i\ (F_o/\partial);\ G/TOP].$$

The subgroup of the right-hand side which is invariant under passage to standard finite covers is $[F_o/\partial;\ G/TOP]$. □

CONCLUDING REMARKS. 1. Note that $[F_o/\partial;\ G/TOP]$ is independent of n. Optimists might conjecture that, for $m = 4$:

$$\mathcal{S}(F^m) \approx [F_o/\partial;\ G/TOP].$$

By the Corollary above this amounts to showing that producting with T^n gives a bijection

$$\mathcal{S}(F) \approx \mathcal{S}_o(T^n \times F).$$

2. It follows from the Proposition above that (cf. [17, p.287]):

$$\mathcal{S}(T^n \times F) \approx \bigoplus_i \binom{n}{i} H^{-i}(F_o,\ \partial;\ G/TOP)$$

The cohomology theory with coefficient spectrum G/TOP has been determined by Sullivan [23]. For example, after tensoring $\mathcal{S}(T^n \times F)$ with Q we have

$$\bigoplus_{k,i} \binom{n}{i} H^{4k-i}(F_o,\ \partial;\ Q).$$

10. PROOF OF THEOREM 4

We will first use Theorem 2 to establish the usual sort of handle lemma which is needed to treat situations as encountered in Theorem 4.

LEMMA 10.1. *Let* $m + n \geq 5$ *and* $\varepsilon > 0$ *be given. There exists a* $\delta > 0$ *so that if* M^{m+n} *and* F^m *are manifolds,* F *compact, and* $f: M \to I^i \times R^{n-i} \times F$ *is a* $p^{-1}(\delta)$*-equivalence over* $I^i \times B_3^{n-i} \times F$ *which is already a homeomorphism from* ∂M *to* $\partial(I^i \times R^{n-i} \times F)$ *(where* p = *projection to* $I^i \times R^{n-i}$*), then there exists an open embedding* $h: f^{-1}(I^i \times \mathring{B}_2^{n-i} \times F) \to I^i \times R^{n-i} \times F$ *which is* $p^{-1}(\varepsilon)$*-homotopic to* $f|\ f^{-1}(I^i \times \mathring{B}_2^{n-i} \times F)$ rel ∂ *provided that* $S_o(T^{n-i} \times I^i \times F) = 0$.

Proof. Let p_1 = projection to R^{n-i} and consider

$$p_1 f: M \to R^{n-i},$$

which is a 2δ-fibration over B_3^{n-i}. By Theorem 1 there is a new map $f': M \to I^i \times R^{n-i} \times F$ for which $p_1 f'$ is an approximate fibration over $B_{2.5}^{n-i}$ and for which $f' \simeq f$ rel ∂M via a homotopy whose projection to R^{n-i} is small.† By identifying $\mathring{B}_{2.5}^{n-i}$ with R^{n-i} it follows that

$$f_1: M_1 = (f')^{-1}(I^i \times \mathring{B}_{2.5}^{n-i} \times F) \xrightarrow{f'} I^i \times \mathring{B}_{2.5}^{n-i} \times F \equiv I^i \times R^{n-i} \times F$$

is a $p_1^{-1}(\gamma)$-equivalence, for every $\gamma > 0$, and is a homeomorphism from ∂M_1 to $\partial(I^i \times R^{n-i} \times F)$. By Theorem 2 there is a homotopy $f_1 \simeq k$ rel ∂M_1, where k is a homeomorphism and the projection of this homotopy to R^{n-i} is small. We may assume that f is given to be a homeomorphism from a boundary collar of ∂M in M to a boundary collar of $\partial(I^i \times R^{n-i} \times F)$ in $I^i \times R^{n-i} \times F$.

---

†Compare with step II in the proof of Theorem 2.

Preserving these boundary collars we have $f_1 = k$ on a boundary collar of ∂M_1 in M_1.

To retrieve the control in the I^i-factor we will have to do some more work. What we do is define a homeomorphism $j: M_1 \to I^i \times \mathring{B}^{n-i}_{2.5} \times F$ by

$$j = (\theta \times \mathrm{id})^{-1} k\tilde{\theta},$$

where $\theta: I^i \times \mathring{B}^{n-i}_{2.5} \to I^i \times \mathring{B}^{n-i}_{2.5}$ is a homeomorphism with compact support which is id on $\partial I^i \times \mathring{B}^{n-i}_{2.5}$, which is id on a small neighborhood of $\{v\} \times \mathring{B}^{n-i}_{2.5}$ (v = center point of I^i), and which radially blows out a slightly larger neighborhood of $\{v\} \times \mathring{B}^{n-i}_{2.5}$ so that it almost contains $I^i \times B^{n-i}_{2.4}$. We can do this so that θ only moves I^i-coordinates. The homeomorphism, $\tilde{\theta}: M_1 \to M_1$, which comes from Theorem 3.6, has compact support and is constructed so that $p(\theta \times \mathrm{id})^{-1} f$ is close to $pf\tilde{\theta}^{-1}$. Because of our assumption that f is already a homeomorphism between boundary collars we can construct $\tilde{\theta}$ so that $f\tilde{\theta} = (\theta \times \mathrm{id})f$ on ∂M_1. Then $h = j| f^{-1}(I^i \times \mathring{B}^{n-i}_2 \times F)$ is easily seen to be our desired open embedding which is $p^{-1}(\varepsilon)$-homotopic to $f| f^{-1}(I^i \times \mathring{B}^{n-i}_2 \times F)$ rel ∂. This construction should be compared with the last paragraph in the proof of Theorem 5.2. □

Proof of Theorem 4. The proof is more-or-less standard. If B has a handle decomposition, then we can use Lemma 10.1 to inductively work through the handles. If B does not have a handle decomposition, then we have to use the more painful chart-by-chart approach. □

APPENDIX 1

As indicated in Section 1 the purpose of this section is to give a proof of the α-Approximation Theorem of [7] based on the engulfing techniques of this paper. This not only acquaints the reader with the definitions and notation which is used in this paper, but it also illustrates the "shuffle" trick, which is the new idea common to the proofs of most all of the theorems in this paper.

α-APPROXIMATION THEOREM [7]. *Let* N^n *be a manifold,* $n \geq 5$, *and let* α *be an open cover of* N. *There exists an open cover* β *of* N *so that if* M^n *is a manifold and* $f\colon M \to N$ *is a* β*-equivalence which is a homeomorphism from* ∂M *to* ∂N, *then* f *is* α*-close to a homeomorphism* $h\colon M \to N$ *for which* $h = f$ *on* ∂M.

As usual any result of this type is deduced from an appropriate handle lemma.

HANDLE LEMMA. *Let* $n \geq 5$ *be an integer and let* $\varepsilon > 0$ *be given. There exists a* $\delta > 0$ *so that if* M^m *is a manifold,* $\partial M = \emptyset$, *and* $f\colon M^n \to R^n$ *is a* δ*-equivalence over* B^n_3, *then there is an open embedding* $g\colon f^{-1}(\mathring{B}^n_2) \to R^n$ *which is* ε*-homotopic to* $f|\ f^{-1}(\mathring{B}^n_2)$.

If the reader cannot guess what it means for f to be a δ-equivalence over B^n_3 he should look at the definition in Section 2. Also we remark that the Handle Lemma is nothing but a special case of Lemma 10.1, and the α-Approximation Theorem is nothing but a special case of Theorem 4 of Section 1. The proofs are also similar.

Proof of the α-Approximation Theorem. We are given a β-equivalence $f\colon M^n \to N^n$. For the moment assume that $\partial M = \emptyset = \partial N$. Write $N = \cup_{i=1}^{\infty} N_i$, where the N_i are copies of R^n and $\{N_i\}_{i=1}^{\infty}$ is a star-finite open cover of N. Choose $C_i \subset N_i$

to be compact so that $\cup_{i=1}^{\infty} C_i = N$ and let $\tilde{C}_i \subset N_i$ be a compactum containing C_i in its interior. Since $N_i \equiv R^n$ we can identify C_i with B_1^n and $\tilde{C}_i$ with B_2^n. By the Handle Lemma there is an open embedding $g_i\colon f^{-1}((\tilde{C}_i)^\circ) \to N_i$ which is close to $f|\ f^{-1}((\tilde{C}_i)^\circ)$. Now use the Deformation Theorem of Edwards-Kirby [10] to adjust the g_i to get open embeddings $h_i\colon f^{-1}(\tilde{C}_i)^\circ) \to N_i$ so that $h_i = h_j$ on $f^{-1}(C_i) \cap f^{-1}(C_j)$. Thus the $h_i|\ f^{-1}(C_i)$ piece together to yield a homeomorphism $h\colon M \to N$ which is close to f.

In the case where $\partial M \neq \emptyset$ and $\partial N \neq \emptyset$ one assumes that f is already a homeomorphism over boundary collars of ∂M and ∂N. Then the above idea is used on the interiors of M and N, and the Deformation Theorem of [10] is again applied to make this homeomorphism compatible with the given homeomorphism on the boundary collars. □

Proof of the Handle Lemma. We are given a proper map $f\colon M^n \to R^n$ which is a δ-equivalence over B_3^n. We regard $T^{n-1} \times R$ as an open subset of B_3^n so that

$$B_2^n = B_2^{n-1} \times [-2,\ 2] \xrightarrow{e^{n-1}\times id} T^{n-1} \times R \hookrightarrow B_3^n$$

is the inclusion, where $e\colon R \to S^1$ is the covering projection defined by $e(x) = e^{\pi i x/4}$ and $T^n = S^1 \times \cdots \times S^1$. By restriction we get a proper map

$$f_o = f|\colon f^{-1}(T^{n-1} \times R) \to T^{n-1} \times R.$$

If δ is small enough, f_o is a δ-equivalence over $T^{n-1} \times [-3,\ 3]$. By engulfing we can wrap this up to obtain a manifold M_1^n and a δ_1-equivalence $f_1\colon M_1 \to T^n$ so that δ_1 is small if δ is small, and for which there is an open embedding $\phi\colon f^{-1}(\mathring{B}_2^n) \to M_1$ making the following diagram commute:

$$\begin{array}{ccc} M_1 & \xrightarrow{f_1} & T^n \\ \uparrow{\scriptstyle\phi} & & \uparrow{\scriptstyle e^n} \\ f^{-1}(\mathring{B}_2^n) & \xrightarrow{f} & \mathring{B}_2^n \end{array}$$

This wrapping up is treated in detail in Lemma 4.1. It should be no problem for anyone who has had experience in this area. Roughly what one does is construct (by engulfing) a homeomorphism $w: f^{-1}(T^{n-1} \times R) \to f^{-1}(T^{n-1} \times R)$ which is supported on $f^{-1}(T^{n-1} \times [-2.5, 2.5])$ and which takes $f^{-1}(T^{n-1} \times \{-2.4\})$ close to the set $f^{-1}(T^{n-1} \times \{2.4\})$. Then we define

$$M_1 = wf^{-1}(T^{n-1} \times (-\infty, -2.4]) - f^{-1}(T^{n-1} \times (-\infty, -2.4))/\sim,$$

where $\sim$ is the equivalence relation generated by $x \sim w(x)$, for $x \in f^{-1}(T^{n-1} \times \{-2.4\})$. If this is done carefully enough we get the δ_1-equivalence $f_1: M_1 \to T^n$ and the open embedding ϕ which makes the diagram commute. We also remark that this is just a simple variation of Siebenmenn's twist-gluing process (twist = identity) [21].

The next step is form the pullback diagram:

$$\begin{array}{ccc} M_2 & \xrightarrow{f_2} & R^n \\ \downarrow & & \downarrow e^n \\ M_1 & \xrightarrow{f_1} & T^n \end{array}$$

Then M_2^n is a manifold and since f_1 is a δ_1-equivalence, f_2 is seen to be a δ_2-equivalence, where δ_2 is small if δ_1 is small. Note that the commutativity of this diagram and the one above give us an open embedding $\psi: f^{-1}(\mathring{B}_2^n) \to M_2$ for which the following diagram commutes:

$$\begin{array}{ccc} M_2 & \xrightarrow{f_2} & R^n \\ \psi \uparrow & & \uparrow \\ f^{-1}(\mathring{B}_2^n) & \xrightarrow{f} & \mathring{B}_2^n \end{array}$$

So to finish the proof of the Handle Lemma we will only have to show that $f_2| f_2^{-1}(\mathring{B}_2^n)$ is ε-close to an open embedding, where $\varepsilon = \varepsilon(\delta_2)$.

By surgery theory the δ_1-equivalence $f_1: M_1 \to T^n$ is homotopic to a homeomorphism $h: M_1 \to T^n$. This lifts to a homeomorphism $\tilde{h}: M_2 \to R^n$, and the homotopy $h \simeq f_1$ lifts to a bounded homotopy $\tilde{h} \simeq f_2$. Choose $K > 0$

large and define $f_3: M_2 \to R^n$ so that $f_3 = f_2$ over B^n_K, $f_3 = \tilde{h}$ over $R^n - \mathring{B}^n_{K+1}$, and over $B^n_{K+1} - \mathring{B}^n_K$ we define f_3 via use of the bounded homotopy $\tilde{h} \simeq f_2$. Now let $\gamma: R^n \to R^n$ be a radially-defined homeomorphism for which $\gamma = \text{id}$ on B^n_2 and $\gamma(B^n_{K+1}) = B^n_3$. Then $f_4 = \gamma f_3: M_2 \to R^n$ is a proper map which equals f_2 over B^n_2, which is a homeomorphism over $R^n - \mathring{B}^n_3$, and which is a δ_2-equivalence provided that K is large enough.

We are now ready to demonstrate the "shuffle" trick which we mentioned earlier. Let $\theta: R^n \to R^n$ be a homeomorphism such that

(1) θ affects only the first coordinate of any point,

(2) θ is supported on B^n_4,

(3) $\theta^{-1}(B^n_2) \cap B^n_{3.5} = \emptyset$.

By engulfing we can "cover" θ with a homeomorphism $\tilde{\theta}: M_2 \to M_2$ so that $\tilde{\theta}$ is supported on $f_4^{-1}(B^n_5)$ and $f_4\tilde{\theta}$ is close to θf_4. This closeness depends on the size of δ_2 above. (See Lemma 3.5 for a proof of the existence of $\tilde{\theta}$.) Now define

$$h = \theta f_4 \tilde{\theta}^{-1} \mid f_4^{-1}(\mathring{B}^n_2): f_4^{-1}(\mathring{B}^n_2) \to R^n.$$

Then $d(h, f_4) = d(\theta f_4 \tilde{\theta}^{-1}, f_4)$ is small, and since f_4 is a homeomorphism over $R^n - \mathring{B}^n_3$ we conclude that h is an open embedding. □

APPENDIX 2

As mentioned in Section 1 the purpose of this section is to apply the results of this paper to prove that a locally homotopically unknotted manifold pair is locally flat.

Recall that a manifold N^n is said to be *locally flat* in a manifold M^{n+k} if N is closed in M and for each $x \in N$ there is a neighborhood $U \subset M$ of x so that the pair $(U, U \cap N)$ is homeomorphic to the pair $((U \cap N) \times R^k, (U \cap N) \times \{0\})$. We say that N is *locally homotopically unknotted* in M if N is closed in M and the following are true:

(1) $k \neq 2$. N is 1-LCC embedded in M, i.e., for each $x \in N$ and neighborhood $U \subset N$ of x, there is a neighborhood $V \subset U$ of x so that each loop in $V - N$ is null-homotopic in $U - N$.

(2) $k = 2$. For each $x \in N$ and neighborhood $U \subset M$ of x, there exists a neighborhood $V \subset U$ of x so that each mapping of a sphere into $V - N$ is null-homotopic in $U - N$ (for 1-spheres we must additionally assume that the mapping is null-homologous in $U - N$).

Clearly if N is locally flat in M, then N is locally homotopically unknotted in M. Here is a converse which is well-known by now.

THEOREM. *If* N^n *is locally homotopically unknotted in* M^{n+k} *and* $n + k \geq 5$, *then* N *is locally flat in* M.

REMARKS. 1. For simplicity we restrict ourselves to manifolds without boundary. There is, of course, the usual version with boundary.

2. The proof of this result which is briefly sketched below enables us to get by with considerably fewer assumptions. Namely (a) the assumption that N

is a manifold can be replaced by the weaker assumption that $N \times R$ is a manifold and (b) the assumption that M is a manifold can be replaced by the weaker assumption that M is a homology manifold whose non-manifold part lies in N.

3. For $k = 1$ this result is due to Cernavskii [3] and Seebeck [19], with an important generalization being obtained by Ferry [14].

4. For $k = 2$ this result is due to Seebeck. In [5] the author gave another proof of the $k = 2$ case which is essentially what is sketched below.

5. For $k \geq 3$ this result is due to Bryant and Seebeck [2].

6. Finally we mention that there are important generalizations of this result which are obtained by Quinn via use of parametrized handlbody theory [18].

Now for a brief sketch of the proof of the Theorem. Since we are interested only in local information we may assume that $N = R^n$. Using duality and a little homotopy theory it is possible to find a closed neighborhood $\tilde{N}$ of N in M and a proper map $f: \tilde{N} - N \to S^{k-1} \times (0, 1] \times R^n$ which is a $p^{-1}(\alpha)$-equivalence over $S^{k-1} \times (0, 1/2] \times R^n$, where p = projection to $(0, 1] \times R^n$ and α is an open cover of $(0, 1] \times R^n$ whose mesh $\to 0$ as we approach $\{0\} \times R^n$. This is essentially the homotopy-theoretic part of the proof given in [5]. By Theorem 1 we can assume that the composition

$$f_1: N_1 = f^{-1}(S^{k-1} \times (0, \varepsilon) \times R^n) \xrightarrow{f} S^{k-1} \times (0, \varepsilon) \times R^n \xrightarrow{p} (0, \varepsilon) \times R^n$$

is an approximate fibration, for some small $\varepsilon > 0$. So by identifying $(0, \varepsilon)$ with R we conclude that $f_1: N_1 \to S^{k-1} \times R^{n+1}$ is a $p^{-1}(\delta)$-equivalence for all δ. By Theorem 2 and the Corollary of Theorem 3 there is a $p^{-1}(\gamma)$-homotopy (γ small) of f_1 to a homeomorphism $h_1: N_1 \to S^{k-1} \times R^{n+1} = S^{k-1} \times (0,\varepsilon) \times R^n$. Thus h extends via the identity to give a homeomorphism of $N_1 \cup N$ with $\mathring{B}^k \times R^n$, where $\mathring{B}^k$ is the open k-ball being thought of as the open cone over S^{k-1}. □

REMARKS. It might be of some interest to point out exactly what calculations are needed from surgery theory for the above proof. For $k = 1$ and $k = 2$ the calculation $\mathcal{S}_o(T^{n+k}) = 0$ is needed. It is possible to deduce this from Edwards' Approximation Theorem and Theorem 1 of this paper. This is explained in the second paragraph following the Corollary of Theorem 1 (in Section 1). For $k \geq 3$ we need the calculation $\mathcal{S}_o(T^{n+1} \times S^{k-1}) = 0$. Is there a proof of this which avoids surgery theory?

APPENDIX 3

The purpose of this section is to show how the techniques of this paper can be used to establish a result on the approximation of approximate fibrations by block bundles.

We will first need the definition of a block bundle. A proper map $p\colon E \to B$ of a manifold to a PL manifold is said to be a *block bundle* with fiber a compact manifold F provided that for each simplex σ in a fixed triangulation of B there is a block preserving homeomorphism $h\colon p^{-1}(\sigma) \to \sigma \times F$. That is, $h(p^{-1}(\tau)) = \tau \times F$ for all faces τ of σ. In the sequel we will refer to the simplices σ of B as *blocks*. It is an easy exercise to show that if $p\colon E \to B$ is a block bundle, then there is a block preserving homotopy of p to an approximate fibration (see Quinn [18]).

Here is the main result of this section. It is concerned with the following question: *When is a proper map of a manifold to a* PL *manifold close to a block bundle (with respect to some fine subdivision of the* PL *manifold)?* We have observed in the above paragraph that a necessary condition for this to happen is that the map must be an α-fibration, for α a fine open cover of the PL manifold.

THEOREM. *Let* B^n *be a* PL *manifold,* $\partial B = \emptyset$, *let* α *be an open cover of* B, *and let* $m \geq 5$ *be an integer. There exists an open cover* β *of* B *so that if* M^{m+n} *is a manifold,* $\partial M = \emptyset$, *and* $f\colon M \to B$ *is a* β*-fibration, then* f *is* α*-homotopic to a block bundle* $p\colon M \to B$ *provided that* $\mathrm{Wh}(T^n \times F)$ *is zero, where* F *is the homotopy fiber of* $f\colon M \to B$.

REMARKS. 1. There is, of course, the usual bounded version in which $f\colon \partial M \to \partial B$ is given to be a block bundle. In this case the conclusion states that f is α-homotopic rel ∂M to a block bundle $p\colon M \to B$ provided that $\mathrm{Wh}(T^n \times F)$ is zero. The proof is the same as the unbounded version.

2. This Theorem is due to Quinn [18], who also showed it to be true if B is only assumed to be a polyhedron. In this case B is necessarily a homology manifold and one can use surgery theory to make our proof work (see the proof of Theorem 3.3.2 of [18]).

3. The Whitehead group condition, $Wh(T^n \times F) = 0$, is not too hard to live with, for it is always satisfied if π_1 of each component of F is free abelian, or more generally, poly Z [12]. Actually the vanishing of the full Whitehead group is not needed. There is a functorial direct sum decomposition,

$$Wh(T^n \times F) = Wh(F) \oplus \sum_{i=1}^{n} \binom{n}{i} \tilde{K}_{1-i}(F) \oplus \text{Nil Terms},$$

where $\tilde{K}_0$ is the reduced projective class group functor and the $\tilde{K}_{1-i}$'s $(i \geq 2)$ are the lower algebraic K-theoretic functors of Bass [1, Chapter XII]. In the proof that we give of the above Theorem the obstructions that we encounter lie only in the subgroup $Wh(F) \oplus \sum_{i=1}^{n} \binom{n}{i} \tilde{K}_{1-i}(F)$ of $Wh(T^n \times F)$.

4. The Whitehead group condition, $Wh(T^n \times F) = 0$, is also easy to apply without dredging up the definition of F. Let $R^n \subset B$ be a copy of euclidean n-space in B. If β is fine enough, then the condition $Wh(T^n \times F) = 0$ is equivalent to the condition that $Wh(T^n \times f^{-1}(B_1^n))$ has 0-image in $Wh(T^n \times f^{-1}(B_2^n))$ under the inclusion-induced map. To see this let $q: E \to B$ be the mapping path fibration of f and let $h: M \to E$ be a homotopy equivalence for which $qh = f$. Since f is a β-fibration, h must be a $q^{-1}(\gamma)$-equivalence by Proposition 2.3. Note that $q^{-1}(B_r^n)$ is homotopy equivalent to F, for any r. Thus we can find maps

$$F \xrightarrow{a} f^{-1}(B_1^n) \xrightarrow{b} F \xrightarrow{c} f^{-1}(B_2^n) \xrightarrow{d} F$$

so that $cb: f^{-1}(B_1^n) \to f^{-1}(B_2^n)$ is homotopic to inclusion and $dcba: F \to F$ is homotopic to id. It then easily follows that $Wh(T^n \times F) = 0$ if and only if $Wh(T^n \times f^{-1}(B_1^n))$ has 0-image in $Wh(T^n \times f^{-1}(B_2^n))$.

The theorem is easily deduced from a Handle Lemma in the usual sort of way. First it will be useful to establish the following result which makes it precise how the Whitehead group condition is used.

LEMMA. *Let* F *be a compact* ANR *for which* $Wh(T^{n-i} \times F) = 0$ *and let* $f\colon (M^{m+n}, \partial M) \to (I^i \times T^{n-i} \times F, \partial I^i \times T^{n-i} \times F)$ *be a homotopy equivalence of pairs such that* $qf\colon \partial M \to \partial I^i \times T^{n-i}$ *is a block bundle, where* $m \geq 5$ *and* q = *projection to* $I^i \times T^{n-i}$. *Then* $qf\colon M \to I^i \times T^{n-i}$ *is homotopic* rel ∂M *to a block bundle.*

Proof. The strategy is to induct on $n - i$. If $n - i = 0$, then $f\colon (M , \partial M) \to (I^n \times F, \partial I^n \times F)$ is a homotopy equivalence and $qf\colon \partial M \to \partial I^n$ is a block bundle. We may assume that F is the fiber of the block bundle $qf\colon \partial M \to \partial I^n$. Write $\partial I^n = J^{n-1} \cup I^{n-1}$, where I^{n-1} is a face of I^n and J^{n-1} is the closure of the complement of I^{n-1} in ∂I^n. Assume that I^{n-1} is a block in ∂I^n and J^{n-1} is a union of blocks in ∂I^n with respect to the given block bundle $qf\colon \partial M \to \partial I^n$. Thus J^{n-1} simplicially collapses to a point. Inducting on the reverse of this collapse it is easy to construct a homeomorphism $h_o\colon f^{-1}(J^{n-1} \times F) \to J^{n-1} \times F$ which is block preserving homotopic to $f|\ f^{-1}(J^{n-1} \times F)$. Since $Wh(F) = 0$ we can use the s-cobordism theorem with ∂ to extend h_o to a homeomorphism $h\colon M \to I^n \times F$. From this it is easy to construct a block bundle of M to I^n which is homotopic rel ∂M to $qf|\partial M$.

Now for the inductive step let $f\colon (M, \partial M) \to (I^i \times T^{n-i} \times F, \partial I^i \times T^{n-i} \times F)$ be a homotopy equivalence of pairs for which $qf\colon \partial M \to \partial I^i \times T^{n-i}$ is a block bundle with fiber F. (If $i = 0$ this condition is vacuous.) Write $T^{n-i} = T^{n-i-1} \times S^1$ and $S^1 = [0, \pi] \cup [\pi, 2\pi]$ with the identification $0 \sim 2\pi$. Since $Wh(T^{n-i} \times F) = 0$ we can use the splitting theorem of [11] to find a homotopy $f \simeq f_1$ rel $\partial\hat{M}$ so that f_1 splits over 0 and π. This means that $M_o = f_1^{-1}(I^i \times T^{n-i-1} \times \{0\} \times F)$ and $M_1 = f_1^{-1}(I^i \times T^{n-i-1} \times \{\pi\} \times F)$ are bicollared submanifolds of M and if

$$M_{[0,\pi]} = f_1^{-1}(I^i \times T^{n-i} \times [0, \pi] \times F),$$

$$M_{[\pi,2\pi]} = f_1^{-1}(I^i \times T^{n-i} \times [\pi,\ 2\pi] \times F),$$

then we have homotopy equivalences

$$f_1|\ M_o: M_o \to I^i \times T^{n-i-1} \times \{0\} \times F,$$

$$f_1|\ M_\pi: M_\pi \to I^i \times T^{n-i-1} \times \{\pi\} \times F,$$

$$f_1|\ M_{[0,\pi]}: M_{[0,\pi]} \to I^i \times T^{n-i-1} \times [0,\ \pi] \times F,$$

$$f_1|\ M_{[\pi,2\pi]}: M_{[\pi,2\pi]} \to I^i \times T^{n-i-1} \times [\pi,\ 2\pi] \times F.$$

Assume that notation is chosen so that $\partial I^i \times T^{n-i-1} \times [0,\ \pi]$ and $\partial I^i \times T^{n-i-1} \times [0,\ \pi]$ are unions of blocks in $\partial I^i \times T^{n-i}$ with respect to the given block bundle $qf: \partial M \to \partial I^i \times T^{n-i}$.

Using the inductive hypothesis we get a homotopy $f_1 \simeq f_2$ rel ∂M so that $qf_2|\ M_o: M_o \to I^i \times T^{n-i-1} \times \{0\}$ and $qf_2|\ M_\pi: M_\pi \to I^i \times T^{n-i-1} \times \{\pi\}$ are block bundles. Now apply the inductive hypothesis to $f_2|\ M_{[0,\pi]}: M_{[0,\pi]} \to I^i \times T^{n-i-1} \times [0,\ \pi] \times F$ and $f_2|\ M_{[\pi,2\pi]}: M_{[\pi,2\pi]} \to I^i \times T^{n-i-1} \times [\pi,\ 2\pi] \times F$ and we have a homotopy $f_2 \simeq f_3$ rel $(\partial M \cup M_o \cup M_1)$ such that $qf_3: M \to I^i \times T^{n-i}$ is a block bundle. □

REMARK. We point out that the above proof implies that if $qf: \partial M \to \partial I^i \times T^{n-i}$ is given to be a block bundle over blocks of small diameter, then the block bundle $p: M \to I^i \times T^{n-i}$ which is constructed above is also over blocks of small diameter.

We are now ready for our Handle Lemma. It is left as an exercise for the reader to show how the Handle Lemma implies the Theorem.

HANDLE LEMMA. *Let* $m \geq 5$ *and* $n \geq 1$ *be integers, and let* $\varepsilon > 0$ *be given. There exists a* $\delta > 0$ *so that if* M^{m+n} *is a manifold and* $f: M^{m+n} \to I^i \times R^{n-i}$ *is a* δ*-fibration over* $I^i \times B_3^{n-i}$ *satisfying* (1) $f|\partial M: \partial M \to \partial I^i \times R^{n-i}$ *is a block bundle over blocks of diameter* $< \delta$ *and* (2) *the inclusion-induced image of* $Wh(T^{n-i} \times f^{-1}(I^i \times B_1^{n-i}))$ *in* $Wh(T^{n-i} \times f^{-1}(I^i \times B_2^{n-i}))$ *is* 0, *then there* *is an* ε*-homotopy* $f \simeq f'$ rel $\partial M \cup f^{-1}(I^i \times (R^{n-i} - \mathring{B}_3^{n-i}))$ *so that* f' *is a block bundle over* $(\partial I^i \times R^{n-i}) \cup (I^i \times B_1^{n-i})$ *and over blocks of diameter* $< \varepsilon$.

Proof. By Theorem 1, and the above remark about deforming block bundles to approximate fibrations, we may assume that f is an approximate fibration. Thus we can find a manifold F (the fiber of $f|\partial M\colon \partial M \to \partial I^i \times R^{n-i}$)† and a homotopy equivalence $\tilde{f}\colon M \to I^i \times R^{n-i} \times F$ so that $p\tilde{f} = f$, where p is projection to $I^i \times R^{n-i}$. By Proposition 2.3, $\tilde{f}$ must be a $p^{-1}(\tilde{\delta})$-equivalence, where $\tilde{\delta}$ is small if δ is small. As in the proof of the Handle Lemma in Appendix 1 let $T^{n-i-1} \times R$ be an open subset of B_3^{n-i} so that the composition

$$B_2^{n-i} = B_2^{n-i-1} \times [-2, 2] \xrightarrow{e^{n-i-1} \times id} T^{n-i-1} \times R \hookrightarrow B_3^{n-i}$$

is the inclusion. By restriction we get a map

$$f_o = \tilde{f}|\tilde{f}^{-1}(I^i \times T^{n-i-1} \times R \times F)\colon \tilde{f}^{-1}(I^i \times T^{n-i-1} \times R \times F) \to I^i \times T^{n-i-1} \times R \times F,$$

which is a $p^{-1}(\tilde{\delta})$-equivalence over $I^i \times T^{n-i-1} \times [-3, 3] \times F$. We can assume that $I^i \times T^{n-i-1} \times [-3, 3]$ is a union of the blocks in $I^i \times R^{n-i}$ over which $f|\partial$ is a block bundle. Also we can assume that for some fine subdivision of $[-3, 3]$, $-3 = t_o < t_1 < \cdots < t_{k-1} < t_k = 3$, each $I^i \times T^{n-i-1} \times [t_{i-1}, t_i]$ is also a union of these blocks.

Now using engulfing we can wrap this up to obtain a manifold M_1^{m+n} and a $q^{-1}(\delta_1)$-equivalence $f_1\colon M_1 \to I^i \times T^{n-i} \times F$ so that

(1) q = projection to $I^i \times T^{n-i}$ and δ_1 is small if $\tilde{\delta}$ is small,

(2) there is an open embedding $\phi\colon \tilde{f}^{-1}(I^i \times \mathring{B}_2^{n-i} \times F) \to M_1$ so that the following diagram commutes:

---

† If $i = 0$, F is the homotopy fiber of f and is no longer presumed to be a manifold. It is shown below that it can be taken to be a manifold.

$$\begin{array}{ccc}
M_1 & \xrightarrow{\quad f_1 \quad} & I^i \times T^{n-i} \times F \\
\Big\uparrow \phi & & \Big\uparrow \mathrm{id} \times e^{n-i} \times \mathrm{id} \\
\tilde{f}^{-1}(I^i \times \mathring{B}_2^{n-i} \times F) & \xrightarrow{\quad f_o \quad} & I^i \times \mathring{B}_2^{n-i} \times F
\end{array}$$

(3) $qf_1\colon \partial M_1 \to \partial I^i \times T^{n-i}$ is a block bundle which extends the block bundle on ∂ which is induced by the commutative diagram above.

As in the wrapping up procedure in the proof of the Handle Lemma in Appendix 1 we are using the idea of Lemma 4.1. We need an engulfing homeomorphism

$$w\colon f_o^{-1}(I^i \times T^{n-i-1} \times R \times F) \to f_o^{-1}(I^i \times T^{n-i-1} \times R \times F)$$

which is supported on $f_o^{-1}(I^i \times T^{n-i-1} \times [-2.5, 2.5] \times F)$ and which takes $f_o^{-1}(I^i \times T^{n-i-1} \times \{-2.4\} \times F)$ close to $f_o^{-1}(I^i \times T^{n-i-1} \times \{2.4\} \times F)$. The problem here is somewhat different because we now have manifolds with boundary. So we first use the s-cobordism theorem to work through the blocks in $\partial I^i \times T^{n-i-1} \times [-3, 3]$ to build the engulfing homeomorphism on the boundary, then we compose this with another engulfing homeomorphism which takes care of the interior. Relative engulfing of this sort is specifically mentioned in the Concluding Remarks following Theorem 3.6. Then, as in Appendix 1, we define

$$M_1 = wf_o^{-1}(I^i \times T^{n-i-1} \times (-\infty, -2.4] \times F) - f_o^{-1}(I^i \times T^{n-i-1} \times (-\infty, -2.4) \times F)/\sim,$$

where $\sim$ is the equivalence relation generated by $x \sim w(x)$, for all $x \in f_o^{-1}(I^i \times T^{n-i-1} \times \{-2.4\} \times F)$. Using the above Lemma on the construction of block bundles we can find a block bundle $q_1\colon M_1 \to I^i \times T^{n-i}$ such that $q_1 \simeq qf_1$ rel ∂M_1. Moreover the blocks in $I^i \times T^{n-i}$ have small diameter. At this point we can now assume that F is a manifold.

Now construct the following pullback diagram:

$$\begin{array}{ccc} M_2 & \xrightarrow{\;f_2\;} & I^i \times R^{n-i} \times F \\ \big\downarrow & & \big\downarrow \scriptstyle{id \times e^{n-i} \times id} \\ M_1 & \xrightarrow{\;f_1\;} & I^i \times T^{n-i} \times F \end{array}$$

Then M_2 is a manifold and f_2 is a $p^{-1}(\delta_2)$-equivalence, where δ_2 is small if δ_1 is small. Also there is a block bundle $q_2: M_2 \to I^i \times R^{n-i}$ which covers q_1 and for which $pf_2|\partial M_2 = q_2$. The homotopy $qf_1 \simeq q_1$ rel ∂M_1 lifts to a bounded homotopy $pf_2 \simeq q_2$ rel ∂M_2. It is easy to construct a bounded homotopy $f_2 \simeq f_3$ rel ∂M_2 so that $q_2 = pf_3$. Thus $f_3: M_2 \to I^i \times R^{n-i} \times F$ is a block preserving homotopy equivalence. Note that there is an infinite collapse of $I^i \times R^{n-i}$ to a point. Working through the reverse of this collapse we can construct a block preserving homeomorphism $h_2: M_2 \to I^i \times R^{n-i} \times F$ which is block preserving homotopic to f_3. Since the blocks are small we can construct a $p^{-1}(\gamma)$-homotopy $f_2 \simeq f_2'$ so that $f_2' = h$ on ∂M_2, where γ is small. Thus $f_2': M_2 \to I^i \times T^{n-i} \times F$ is a $p^{-1}(\delta_2')$-equivalence which is boundedly homotopic to the homeomorphism h rel ∂M_2.

Just as in the proof of the Handle Lemma in Appendix 1 we can use a "shuffle" trick to construct an open embedding $g: (f_2')^{-1}(I^i \times \mathring{B}_2^{n-i} \times F) \to I^i \times R^{n-i} \times F$ such that $g = h$ on ∂ and such that g is $p^{-1}(\varepsilon_1)$-homotopic to $f_2'|$ rel ∂ , where ε_1 is small.† The commuting diagrams give us an open embedding $\psi: (\tilde{f})^{-1}(I^i \times \mathring{B}_2^{n-i} \times F) \to M_2$ so that $f_2\psi = \tilde{f}$ over $I^i \times \mathring{B}_2^{n-2} \times F$. Thus $\psi^{-1}g^{-1}(I^i \times B_1^{n-i} \times F) \xrightarrow{pg\psi} I^i \times B_1^{n-i}$ is a block bundle where $\psi^{-1}g^{-1}(I^i \times B_1^{n-i} \times F)$ is now in M. We may assume in our given block bundle $f|\partial M: \partial M \to \partial I^i \times R^{n-i}$, that $\partial I^i \times B_1^{n-i}$ is a union of blocks. Thus $\psi^{-1}g^{-1}(\partial I^i \times B_1^{n-i}) = f^{-1}(\partial I^i \times B_1^{n-i})$ and we can find a block preserving homotopy of $pg\psi$ to $p_1: \psi^{-1}g^{-1}(I^i \times B_1^{n-i} \times F) \to I^i \times B_1^{n-i}$ so that $p_1 = f$

---

† The "shuffle" trick must now be done rel ∂. It is in the same spirit as the unbounded case, but more care must be taken. This is done in the proof of Theorem 2.

on $f^{-1}(\partial I^i \times B_1^{n-i})$. It is clear that there is a small homotopy of f to f' rel ∂M so that $f' = p_1$ over $I^i \times B_1^{n-i}$. $\square$

REFERENCES

1. H. Bass, *Algebraic K-theory*, W. A. Benjamin, New York, 1968.

2. J. L. Bryant and C. L. Seebeck, III, *Locally nice embeddings in codimension three*, Quart. J. Math. Oxford Ser.(2) 21(1970), 265-272.

3. A. V. Cernavskii, *Coincidence of local flatness and local simple connectedness for embeddings of* (n-1)*-dimensional manifolds in* n *dimensional manifolds when* n > 4, Mat. Sb. 91(133) (1973), 279-286.

4. T. A. Chapman, *Approximating maps into fiber bundles by homeomorphisms*, to appear in Rocky Mountain J. of Math.

5. ____________, *Locally homotopically unknotted embeddings of manifolds in codimension two are locally flat*, to appear in Topology.

6. ____________, *Approximation results in Hilbert cube manifolds*, preprint.

7. T. A. Chapman and Steve Ferry, *Approximating homotopy equivalences by homeomorphisms*, to appear in American J. of Math.

8. D. Coram and P. Duvall, *Approximate fibrations*, Rocky Mountain J. of Math. 7(1977), 275-288.

9. R. D. Edwards, *Approximating certain cell-like maps by homeomorphisms*, Annals of Math., to appear.

10. R. D. Edwards and R. C. Kirby, *Deformations of spaces of imbeddings*, Annals of Math. 93(1971), 63-88.

11. F. T. Farrell, *The obstruction to fibering a manifold over the circle*, Bull. A. M. S. 73(1967), 737-740.

12. F. T. Farrell and W. C. Hsiang, *A formula for* $K_1(R_\alpha[T])$, Proc. Symp. Pure Math. 17, Applications of Categorical Algebra, AMS, 1970.

13. Steve Ferry, *Approximate fibrations with non-finite fibers*, P. A. M. S. 64(1977), 335-345.

14. ____________, *Homotoping* ε*-maps to homeomorphisms*, to appear in American J. Math.

15. ____________, *Approximate fibrations over* S^1, in preparation.

16. J. F. P. Hudson, *Piecewise linear topology*, Benjamin, New York, 1969.

17. R. Kirby and L. Siebenmann, *Foundational essays on topological manifolds, smoothings, and triangulations*, Princeton University Press, Princeton, N. J., 1977.

18. Frank Quinn, *Ends of maps, and applications*, preprint.

19. C. L. Seebeck, III, *Codimension one manifolds that are locally homotopically unknotted on one side,* I and II, preprint.

20. L. C. Siebenmann, *Approximating cellular maps by homeomorphisms,* Topology 11(1972), 271-294.

21. ________________, *A total Whitehead torsion obstruction,* Comment. Math. Helv. 45(1970), 1-48.

22. L. Siebenmann, L. Guillou, H. Hähl, *Les voisinages ourvets réguliers: critères homotopiques d'existence,* Ann. scient. Ec. Norm. Sup. 7(1974), 431-462.

23. D. Sullivan, *Triangulating homotopy equivalences,* Ph.D. Thesis, Princeton, 1966.

Department of Mathematics
University of Kentucky
Lexington, Kentucky 40506

ABCDEFGHIJ-AMS-8987654321

General instructions to authors for
PREPARING REPRODUCTION COPY FOR MEMOIRS

For more detailed instructions send for AMS booklet, "A Guide for Authors of Memoirs." Write to Editorial Offices, American Mathematical Society, P. O. Box 6248, Providence, R. I. 02940.

MEMOIRS are printed by photo-offset from camera copy fully prepared by the author. This means that, except for a reduction in size of 20 to 30%, the finished book will look exactly like the copy submitted. Thus the author will want to use a good quality typewriter with a new, medium-inked black ribbon, and submit clean copy on the appropriate model paper.

Model Paper, provided at no cost by the AMS, is paper marked with blue lines that confine the copy to the appropriate size. Author should specify, when ordering, whether typewriter to be used has PICA-size (10 characters to the inch) or ELITE-size type (12 characters to the inch).

Line Spacing – For best appearance, and economy, a typewriter equipped with a half-space ratchet – 12 notches to the inch – should be used. (This may be purchased and attached at small cost.) Three notches make the desired spacing, which is equivalent to 1-1/2 ordinary single spaces. Where copy has a great many subscripts and superscripts, however, double spacing should be used.

Special Characters may be filled in carefully freehand, using dense black ink, or INSTANT ("rub-on") LETTERING may be used. AMS has a sheet of several hundred most-used symbols and letters which may be purchased for $5.

Diagrams may be drawn in black ink either directly on the model sheet, or on a separate sheet and pasted with rubber cement into spaces left for them in the text. Ballpoint pen is *not* acceptable.

Page Headings (Running Heads) should be centered, in CAPITAL LETTERS (preferably), at the top of the page – just above the blue line and touching it.

LEFT-hand, EVEN-numbered pages should be headed with the AUTHOR'S NAME;

RIGHT-hand, ODD-numbered pages should be headed with the TITLE of the paper (in shortened form if necessary).

Exceptions: PAGE 1 and any other page that carries a display title require NO RUNNING HEADS.

Page Numbers should be at the top of the page, on the same line with the running heads.

LEFT-hand, EVEN numbers – flush with left margin;

RIGHT-hand, ODD numbers – flush with right margin.

Exceptions: PAGE 1 and any other page that carries a display title should have page number, centered below the text, on blue line provided.

FRONT MATTER PAGES should be numbered with Roman numerals (lower case), positioned below text in same manner as described above.

MEMOIRS FORMAT

It is suggested that the material be arranged in pages as indicated below.
Note: Starred items (*) are requirements of publication.

Front Matter (first pages in book, preceding main body of text).

Page i – *Title, *Author's name.

Page iii – Table of contents.

Page iv – *Abstract (at least 1 sentence and at most 300 words).

*1980 Mathematics Subject Classifications represent the primary and secondary subjects of the paper. For the classification scheme, see Annual Subject Indexes of MATHEMATICAL REVIEWS beginning in December 1978.

Key words and phrases, if desired. (A list which covers the content of the paper adequately enough to be useful for an information retrieval system.)

Page v, etc. – Preface, introduction, or any other matter not belonging in body of text.

Page 1 – Chapter Title (dropped 1 inch from top line, and centered).

Beginning of Text.

Footnotes: *Received by the editor date.

Support information – grants, credits, etc.

Last Page (at bottom) – Author's affiliation.